Source Book of Experimental Techniques and Analysis of Polymeric Materials

Susmita Dey Sadhu · Bhasha Sharma ·
Vijay Chaudhary · Anup K. Ghosh

Source Book of Experimental Techniques and Analysis of Polymeric Materials

Volume 1: Synthesis and Identification of Polymers and Allied Materials

Springer

Susmita Dey Sadhu
Department of Polymer Science
Bhaskaracharya College of Applied Sciences
University of Delhi
New Delhi, Delhi, India

Bhasha Sharma
Department of Chemistry
Bhagini Nivedita College
University of Delhi
New Delhi, Delhi, India

Vijay Chaudhary
Department of Mechanical Engineering
Amity University Uttar Pradesh
Noida, Uttar Pradesh, India

Anup K. Ghosh
Center for Polymer Science and Engineering
Indian Institute of Technology Delhi
New Delhi, Delhi, India

ISBN 978-981-92-2087-8 ISBN 978-981-92-2088-5 (eBook)
https://doi.org/10.1007/978-981-92-2088-5

This Springer imprint is published by the registered company Springer Nature Singapore Pte Ltd.
The registered company address is: 152 Beach Road, #21-01/04 Gateway East, Singapore 189721, Singapore

Foreword

The field of polymer science continues to drive advancements that shape our modern world. This is a practical subject and an effective learning is guaranteed only through experiments. In this context, **Source Book of Experimental Techniques and Analysis of Polymeric Materials, Volume 1: Synthesis and Identification of Polymers and Allied Materials** serves as an essential resource for students, educators, and early-career researchers who seek a hands-on understanding of polymer chemistry. This book aims to provide a complete solution to the polymer chemists with basics of polymer types, its identification techniques, various types of synthesis methodologies (for thermosets, thermoplastics, and conducting polymers). Industries dealing in similar will also be benefitted by these experiments. In my knowledge, this book has covered the synthesis of polymers along with identification in a unique and extensive manner.

This book is more than a manual of experiments—it is a guide to the foundational techniques and practical wisdom that underpin the synthesis, characterization, and identification of polymers. Each chapter is carefully crafted to provide not only procedural clarity but also theoretical context, enabling readers to grasp the underlying principles of each experiment. Whether it is a step-growth or chain-growth polymerization, spectroscopy or solubility testing, the methodologies presented here reflect both academic rigor and industrial relevance.

As we continue to navigate the challenges of sustainability, materials innovation, and performance optimization, there has never been a greater need for polymer scientists with both conceptual understanding and laboratory competence. This source book responds to that need by fostering both inquiry and skill.

I commend the authors for their meticulous work in assembling this practical compendium. It is a timely contribution to polymer education and will undoubtedly inspire and equip a new generation of material scientists to explore the vast and versatile world of polymers. The book is relevant to today's requirement of polymer science fraternity both industries and academics.

Anil K. Bhowmick
Department of Chemical
and Biomolecular Engineering
The University of Houston
Houston, TX, USA

Preamble

Polymer materials have become cutting-edge in modern research and industrial development, dominating myriad applications from consumer goods to state-of-the-art technologies. The synthesis of polymers is a dynamic field that demands a comprehensive understanding of material properties, chemical processes, and molecular architecture, spanning from initial molecular design to large-scale production. Identification and characterization of such materials are similarly significant which promises to meet detailed functional and performance requirements in industries such as electronics, energy, and healthcare.

The book has been designed to cover "Molecules to manufacturing to re-manufacturing" in a series of four volumes describing a circular process involving the creation of new products from molecules (Volume 1), their subsequent use in manufacturing (Volume 2), testing and characterising them (Volume 3) and then the recycling or reprocessing of those products back into usable materials (re-manufacturing) to create new products (Volume 4) as below:

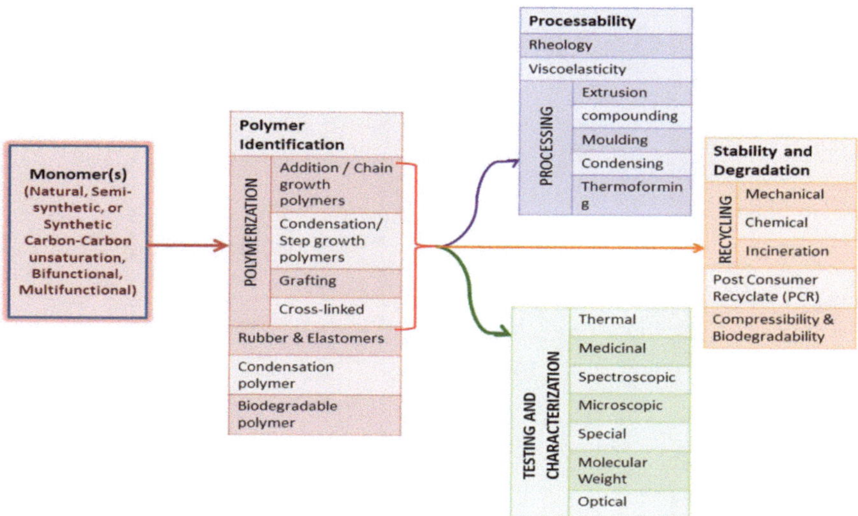

This volume, *Synthesis and Identification of Polymers and Allied Materials*, presents a broad overview of processes involved in developing and analysing polymers. It revolves around the principles of synthesis of polymers, various methods, and techniques applied for their characterization and identification. The content focuses on the synthesis of polymers which intends to empower researchers, students, and professionals with the tools and knowledge that play integral role in innovating the domain of polymer science.

Owing to the increasing demand of high performance and sustainable materials, this volume addresses the innovations in the synthesis of eco-friendly polymers and techniques utilized to configure polymers for specialized industrial applications. The methods mentioned here bestow valuable insights into the development of new polymeric materials covering a wide range of topics, from the basic principles of polymer synthesis and chemical structures to sophisticated testing methodologies. The content is expected to equip readers with the knowledge necessary to design, synthesize, and characterize new polymers; to bridge the gap between academia and industry as well as beneficial for industrial application.

Competing Interests The authors have no competing interests to declare that are relevant to the content of this manuscript.

Introduction to Polymeric Materials—Exploring the Versatility and Their Applications

The unique properties and versatility of polymers make them interesting to study and explore. Polymers are found to have a wide range in properties depending on their chemical nature, glass transition temperature, crystallinity, and so on. These materials find a wide range of applications depending on their nature and properties.

The mechanism and condition of polymer synthesis decides the chemical structure like branching, crystallinity and molecular weight of the polymers. Hence, the chemistry behind the Polymer synthesis is very important to tailor the polymer structure.

Polymers play a crucial role today's manufacturing and industrial applications, for example in packaging, defence, footwear, clothing, automobile etc. as well as in medicine and healthcare. The journey of polymers is full of new technologies, innovations, advancement in applications but also poses environmental challenges that need to be addressed.

A lot of work is being done in combating the environmental issues. Artificial Intelligence in Polymers opens a new horizon in polymer synthesis, processing, and in managing waste.

Polymer since its inception has travelled a long way of evolution. The polymers like starch, cellulose, casein, and other macromolecules were known to mankind long ago. But the concept of polymer started with the early work on modifying natural materials like cellulose, with notable advancements like celluloid in the 1860s by John Wesley Hyatt. He modified cellulose to cellulose nitrate to be used as a substitute of Ivory [1]. In 1830s Cellulose Acetate was developed by Henri Braconnot [2] to create new semi-synthetic materials.

In 1844 Charles Goodyear [3] vulcanized natural rubber which impacted the industry significantly. However, the first synthetic polymer, was created by Leo Baekeland in 1907 in the form of Bakelite [4]. First Hermann Staudinger (his work between 1926 and 1956) proposed the concept of polymers as long chains of atoms held together by covalent bonds [5]. This groundbreaking idea revolutionized the field of polymer science.

Wallace Carothers (1896–1937) developed nylon as a Landmark work in polymer science. His research confirmed the existence of high molecular weight molecules leading to the creation of nylon [6].

The discovery of polypropylene and the development of a new high-density polyethylene process at Phillips Petroleum Company were also recognized as a National Historic Chemical Landmark [7].

Applications of polymers: Polymers can be both man-made and naturally occurring. They are found in a wide range of products, from plastics and composites to clothing, toys, and construction materials. Polymers can be pliable, rigid, elastic, or fibrous, depending on their composition and molecular structure. They are composed of monomers, which are smaller molecules that are linked together to form the long chains characteristic of polymers. Polymers have revolutionized modern society and have countless applications in various industries.

1. Packaging
Polymers are extensively used in packaging materials such as plastic films, containers, bottles, and bags. They offer lightweight, durable, and cost-effective solutions for protecting and preserving food, beverages, pharmaceuticals, and consumer goods.

2. Automotive
Polymers find numerous applications in the automotive industry, including interior components, such as dashboards, seats, and door panels, as well as exterior parts like bumpers, body panels, and tires. They contribute to weight reduction, improved fuel efficiency, and enhanced design flexibility.

3. Construction
Polymers are used in construction for a variety of purposes, including pipes, fittings, insulation materials, roofing membranes, sealants, adhesives, and coatings. They provide durability, energy efficiency, corrosion resistance, and ease of installation.

4. Electronics and Electrical
Polymers play a vital role in the electronics and electrical industry, where they are used for the insulation of wires and cables, printed circuit boards (PCBs), connectors, housings for electronic devices, and displays. They offer electrical insulation properties, lightweight design, and protection against moisture and environmental factors.

5. Medical and Healthcare
Polymers are extensively used in medical and healthcare applications, including surgical instruments, implants, drug delivery systems, wound dressings, prosthetics, and dental materials.

Biocompatible polymers are employed to ensure compatibility with living tissues.

6. Textiles and Fibers
Polymers are utilized in the production of synthetic fibers for textiles, including polyester, nylon, and acrylic. These fibers offer a wide range of properties, such as strength, elasticity, wrinkle resistance, and moisture absorption.

7. Consumer Goods
Polymers are found in a wide range of consumer goods, including household appliances, furniture, toys, sports equipment, and personal care products. They provide versatility in design, cost-effectiveness, and enhanced functionality.

8. Agriculture
Polymers are used in agriculture for various purposes, such as mulching films, greenhouse covers, irrigation systems, crop protection materials, and seed coatings. They help improve crop yields, conserve water, and protect plants from pests and diseases.

9. Energy
Polymers are employed in energy-related applications, including solar panels, wind turbine blades, fuel cells, batteries, and insulation materials for energy-efficient buildings. They contribute to renewable energy generation, energy storage, and improved energy efficiency.

10. Aerospace
Polymers are utilized in the aerospace industry for lightweight components, such as aircraft interiors, structural parts, seals, gaskets, and thermal insulation materials. They help reduce weight, increase fuel efficiency, and meet stringent safety requirements.

Benefits of Polymer

Polymers offer several benefits, which contribute to their widespread use in various industries.

Here are some of the advantages of polymers:

1. Versatility: Polymers can be tailored to have a wide range of properties, allowing for versatility in applications. They can be engineered to be rigid or flexible, transparent or opaque, and have varying degrees of hardness or softness.
2. Lightweight: Many polymers have low density, making them lightweight materials. This property is particularly advantageous in industries such as aerospace, automotive, and packaging, where weight reduction is desirable.
3. Durability: Polymers often exhibit excellent durability and resistance to wear, chemicals, and environmental factors. They can withstand harsh conditions, making them suitable for long-lasting applications.
4. Cost-Effective: Polymers can be produced in large quantities at relatively low costs compared to other materials such as metals or ceramics. This cost-effectiveness makes them economically attractive for various industries.

5. Processability: Polymers can be easily processed into complex shapes using techniques like injection molding, extrusion, and 3D printing. This enables efficient and cost-effective manufacturing of polymer-based products.
6. Insulation Properties: Many polymers possess excellent electrical insulation properties, making them valuable in the electrical and electronics industries. They can be used for insulation in wires, cables, and electronic components.
7. Corrosion Resistance: Certain polymers, such as polypropylene and polyethylene, exhibit high corrosion resistance, making them suitable for applications in corrosive environments or as protective coatings.
8. Design Flexibility: Polymers offer design flexibility, allowing for the creation of intricate and customized shapes. They can be easily molded or fabricated into desired forms, offering design freedom to engineers and designers.
9. Recycling Potential: Many polymers are recyclable, providing opportunities for reducing waste and promoting sustainability. Recycling processes can help recover valuable resources from used polymers and reduce environmental impact.
10. Biocompatibility: Some polymers are biocompatible, meaning they are compatible with living tissues and can be used in medical and healthcare applications such as implants, drug delivery systems, and wound dressings. Polymers and structures: A polymer is a material consisting of very large molecules called macromolecules, composed of many repeating subunits. These monomers are small molecules that can bond together through various chemical reactions to form long chains or networks, which are characteristic of polymers. The specific composition and arrangement of the monomers determine the properties of the polymer. In some cases, polymers can also contain additional components such as additives, fillers, or reinforcing agents to enhance their performance or tailor their properties for specific applications. The versatility of the polymers arises from the ability to customize their composition, molecular structure, and processing techniques to achieve a wide range of desired characteristics.

Long polymer molecules intertwine or "entangle" with each other, just like spaghetti on a plate. These physical entanglements allow polymer solutions to bear some mechanical stress. The molecules are usually coiled up, but when stress is imposed on them they can straighten out somewhat. When the stress is released, they coil up again. This coiling process gives a polymer its elasticity. All long polymer molecules in a sufficiently concentrated solution will have physical entanglements that lead to viscoelastic behavior.

How many is "many?"

This question is difficult since there is no absolute answer. Rather, it depends on what property or characteristic is of interest. For example, one of the important characteristics that defines the behavior of a molten polymer is the tendency for molecules to become entangled with one another, a phenomenon that dramatically affects the flow characteristics, or mechanical properties of the melt. Below a critical point, termed the entanglement molecular weight, the viscosity of the material only increases slowly as the molecules get longer; above this critical threshold, the viscosity increases dramatically as the system becomes increasingly entangled as the molecules get longer. The most important "bulk commodity" polymers–polyethylene and polypropylene–are both materials that include a significant crystalline fraction (40–90%). These crystals contain molecules which, in the ideal case, repeatedly fold back upon themselves. This is characteristic of polymers and, therefore, one can argue this tendency to fold defines a polymer; in polyethylene, the onset of chain folding occurs for chains containing more than about 150 carbon atoms. Systems which do not exhibit polymeric characteristics such as those examples described above are commonly termed oligomers.

In general, polymer contains a variety of molecular forms which may differ from one another in terms of:

- molecular length
- number of branches or side groups
- length of branches
- distance between branches or side groups
- average number of branches or side groups per molecule.

All of these can apply even to one single material–say, polyethylene. Therefore, it is critically important when thinking of polyethylene that this variability is remembered and knowledge base is created for different grades of polymer.

Classifications of polymers: With the evolution of new materials and composites, today we are in an age where a wide variety is available. These wide variety of polymeric materials can be catagorized into many ways. Conventionally, these materials are divided as thermoplastic, thermoset and thermoplastic elastomers based on their behavior. Thermoplastic materials are those materials which flow on heating while they takes a solid structure on cooling. Most of the plastic materials like PP, LDPE, PS, LLDPE, PET, Nylon etc. fall in this category. Thermosets are the materials which once take a shape that cannot be melted and reshaped. It does not flow on heating as it has a 3D network structure. Thermoplastic elastomers are the materials which show the properties of both. It has the high impact strength of elastomers but at the same time can flow on heating and can be reshaped.

Based on applications it can be divided as fiber, rubber, film, plastic etc. Based on origin it may be catagorized as natural plastic (bioplastics), semi-synthetic plastic, and synthetic plastic. Depending on behavior, it is also divided as biodegradable, biocompatible, and nonbiodegradable.

In this section we would like to classify polymeric materials also as

1. Petroleum-based polymer
2. Bio-based, biodegradable and biocompostable Polymers
3. Polymers recyclate

1. Petroleum-Based Polymers:

Petroleum-based polymers are those polymers which are synthesized directly or indirectly from petroleum byproducts. Most of the conventional polymers/synthetic polymers fall in this category. These polymers are chemically synthesized from the hydrocarbons obtained from natural gas, coal, and crude oil or other petroleum

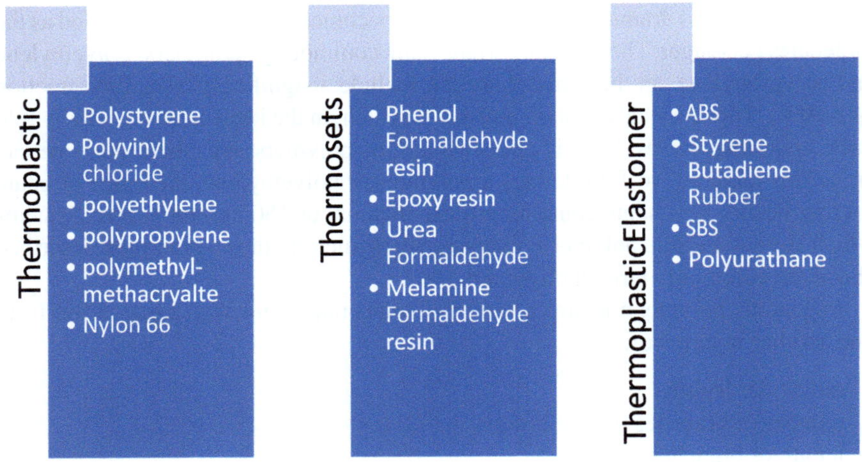

Fig. 1 Classification of petroleum-based polymers

products. Synthetic plastic has vast applications due to its excellent characteristic properties (Luckachan & Pillai 2011) [8]. By changing various reaction parameters, synthetic polymers can be synthesized with a variety of structure (linear, branched, spherical etc.) along with a range of molecular weight. This wide variety gives an option to choose a material with required properties, processibility, and application. Plastic finds applications starting from the domestic to the industrial level products.

Plastics are strong, lightweight, and cheap. These polymers have high molecular weight and consist of hundreds to thousands of repeating units of monomers (Kuhn et al. 2007) [9]. Synthetic plastics are produced by refining crude oil. This distillation process separates high molecular weight crude oil into lighter constituents. Every lighter constituents has a specific hydrocarbon chain of varying sizes and structures for example, Naphtha. The polymerization process releases gases and other toxic materials into the atmosphere. The polymerization parameters decide the polymer structure and molecular weight. The final application is determined based on the characteristic properties of plastic.

Petroleum-based polymers include both thermoplastic, thermosets, and thermoplastic elastomers.

1.1 Commonly Used Synthetic Plastics

- Polyethylene (PE): It is the most commonly used plastic. The ethylene is polymerized to get PE. The first polyethylene was produced in 1934 [10]. PE is tough, has low capacity of moisture adsorption, chemically inert, and is a bad conductor of electricity. The properties depend on the type of branching and crystal structure [11] they have. PEs are used in making electric wires and cables, pipes, medical devices, and automotive and space applications [12–15].
- PE can be of three types-(1) HDPE (High-density polyethylene), (2) LDPE (Low-density polyethylene), and (3) MDPE (Medium-density polyethylene). LDPE is

ductile, flexible, transparent, tough, and has low crystallinity due to high chain branching. Its long chain irregular packing contributes to low crystallinity. LDPE shows tensile strength lower than HDPE and finds applications in preparing containers, plastic films, and bags etc. HDPE is the most versatile plastic material with the high degree of crystallinity. It is highly stable due to the presence of many short branches. HDPE makes milk jugs, water pipes, and detergent bottles [16]. It has a higher melting point than LDPE [17]. MDPE is in between LDPE and HDPE. It shows better-cracking resistance than HDPE but falls in between HDPE and LDPE in hardness.

- Polypropylene (PP): PP is obtained from propylene. This plastic exhibits good mechanical properties at affordable cost. PP can be used in household appliances, packaging, electronics, bottle tubes, filaments, and bags etc. Depending on the molecular arrangements there are three kinds of PP available:
- In atactic PP the methyl groups are arranged in a random manner resulting into low melting point.
- In syntactic PP, the methyl groups are arranged in an alternate fashion. The polymer is semi-crystalline and more homogenous than atactic PP.
- Isotactic PP shows a high degree of crystallinity due to uniform organization of the methyl group. This PP is homogenous and highly stable.
- Polyvinyl Chloride (PVC): PVC is obtained from vinyl chloride. Due to its unstable nature generally it is processed along with stabilizers. Plasticizer is added to it to make it flexible. It is a low cost and durable material.
- Polystyrene (PS): Polystyrene made up of styrene monomer is a rigid, lightweight, and moldable plastic. When polystyrene is heated with blowing air, EPS/ Styrofoam (foam made out of PS) is formed. Polystyrene finds application in making furniture, glasses, and utensils.
- Polyethylene Terepthalate (PET): Polyethylene terephthalate is a copolymer of terepthalic acid and ethylene glycol. It is known for its strength, transparency, and barrier properties, making it suitable for beverage bottles, food containers, and clothing.
- Nylon (PA): Polyamides are a class of polymers that contain amide bonds (-CO-NH-) in their molecular structure. They are known for their strength, durability, and versatility, and are widely used in various applications including textiles, automotive parts, and engineering components. Nylon 6, Nylon 66, Nylon 11 and Nylon 6,10 are well-known examples of polyamides.
- Polytetrafluoroethylene (PTFE): PTFE was first prepared in 1938 by Dupont [18, 19] by polymerizing the monomers tetrafluoroethylene. It is known as teflon and is stable, strong, and resistant to high temperatures. It is used to make cookware and waterproof coating [20].

1.2 Polymerization Techniques: The process of creating plastics is known as polymerization. This process entails connecting numerous monomers to produce a polymer molecule. Monomers are small molecular units that combine through chemical reactions to create large molecules with high molecular weight, referred to as polymers. Various polymerization techniques can be employed to generate different types of plastic materials, including:

A. Condensation Polymerization or step-growth polymerization is a chemical polymerization reaction, where a low molecular weight by-product, known as a condensate (e.g., NH_3, H_2O, HCl), is released while the polymer is formed.
B. Addition Polymerization, also known as Chain Polymerization. Here, the polymer is prepared by combining two or more monomers without any by-product formed. This process is associated with a chain reaction where the monomer molecules are linked with each other.
C. Polymerization by Ionizing Radiation Ionizes to generate free radicals, which start the polymerization process. In ionizing radiation polymerization both solid and liquid mediums are used. Temperature does not impact this process. This process may even continue after taking away the radiation source, and this phenomenon is known as post-polymerization.
D. Non-ionizing Radiation Polymerization (NIRP) is a special type of chain-growth polymerization. The presence of visible or UV light impacts the process. During this process, the reactant monomer absorbs the light directly or through a photosensitizer (which helps to transfer its energy to the monomer).

1.3 Disadvantages of Petroleum-Based Plastics: Plastic is derived from non-renewable energy sources and is not biodegradable. The chemical composition of plastic is resistant to alteration by microorganisms, which contributes to its persistence in the environment. Factors such as high molecular weight, significant crystallinity, and elevated hydrophobicity hinder the degradation process of plastic. The production and disposal of plastic continue to rise annually. Due to its nonbiodegradable nature, effective disposal methods for plastic are lacking. The accumulation of plastic waste has severely impacted the environment, leading to pollution of land, water, and air, and it plays a role in global warming. Additionally, it poses risks to both human and animal health by contributing to infectious and chronic diseases. A substantial volume of plastic waste can also impede groundwater movement. Plastic waste often contains heavy metals, which can degrade soil quality and contaminate water sources. Additionally, plastic negatively impacts soil fertility. The incineration and open burning of plastic waste release dioxins and harmful gases into the atmosphere. The scarcity of appropriate landfill sites further complicates waste management. Accumulated plastic debris in marine environments raises aesthetic issues and can detrimentally affect tourism. Moreover, significant amounts of plastic waste interfere with fishing operations, as debris often becomes entangled in fishing nets alongside captured fish. The ingestion of plastic by aquatic organisms poses serious health risks. This issue extends to other animals and humans, as the consumption of plastic by marine species can introduce harmful substances into the food chain, facilitating the transfer of toxins between organisms. Another critical concern is the presence of microplastics, which are tiny plastic fragments measuring between 0.1 mm and 5 mm in length. Numerous studies have documented the occurrence of microplastics in plankton, fish, and humans. Toxic additives such as phthalates and Bisphenol A (BPA) are incorporated during plastic production and can bioaccumulate in living organisms. While recycling presents a viable solution for reducing plastic waste, it often proves to be less cost-effective, as the production of new plastic can sometimes be cheaper than recycling efforts.

2. Bioplastic, Biodegradable Plastic, and Bio-compostable Plastic

Biodegradable plastics or bioplastics offer a promising solution to traditional plastics by reducing waste, minimizing pollution, and contributing to a more sustainable future. There are different types of biodegradable plastics. All biodegradable plastics are not biocompostable. They need specific conditions to undergo degradation. Most of the biodegradable plastics are obtained from biomass making it a sustainable product.

Biodegradable plastics are the plastics which can be broken down by microorganisms to produce carbon-di-oxide, water and energy along with biomass. Biodegradable plastics are being used to replace conventional plastics in packaging applications, fiber, agricultural and horticultural sectors and making disposable containers etc. The commonly known biodegradable plastics are polylactic acid (PLA), Polyhydroxyalkanoates (PHAs), starch-based polymers, soy proteins and many more. All these polymers are naturally obtained. Although the demand for biodegradable polymers is increasing, the acceptance of biodegradable plastics is likely to depend on factors like

(1) Buying cost and affordability of customers
(2) Rules and regulations by Governments
(3) The performance of the biodegradability and other properties

There are several plastics which are obtained from petroleum sources but they are biodegradable in nature. We can call these biodegradable but not bioplastics.

Therefore, there are three types of biodegradable/biomass-based plastics:

2.1 Bioplastics, whose components are derived from renewable raw materials: The most environment friendly plastic is the one which is both biomass derived and biodegradable. Polylactic acid (PLA), starch and cellulose-based biodegradable plastics and polyhydroxyalkanoate (PHA) are some of the examples.
2.2 Petrochemical-based biodegradable plastics: These plastics are derived from petroleum but are biodegradable. Polybutylene succinate (PBS), poly(butylene adipate-co-terephthalate) (PBAT), and polyvinyl alcohol (PVA) are members of this category.
2.3 This class of bioplastics are derived from biomass but are not biodegradable. For example biomass derived polyethylene, used in shopping bags. Owing to their biomass origin these plastics can be regarded as environment friendly in terms of reduced CO_2 emissions.

2.1 Most widely use biodegradable biomass derived plastics are:

Polylactic acid (PLA):

PLA is a biomass derived biodegradable plastics made from corn, sugarcane, and starch etc. L-lactic acid is prepared through fermentation process from the glucose obtained by the enzyme induced hydrolysis of starch. Finally, PLA is obtained by polymerization of lactic acid. In this preparation method a nutritious food source is used as the raw material for polymerization. This adds to the cost of the polymer. Scientists are trying to prepare PLA from non-edible biomass like cellulose. The properties of PLA like crystalline structure, mechanical strength, melting point, and biodegeneration of PLA depend largely on the quantity of its stereo-isomers like D-lactic acid, L-lactic acid, and DL-lactic acid present in the structure. PLA does not undergo decomposition in the normal condition with slow biodegradability but it biodegrades in composting conditions within six months. PLA is used in packaging, wrappings, lamination, making of containers (bottles and cups), etc. It is used as bone fixation screws, suture threads, drug delivery devices etc. in the biomedical field.

Polyhydroxyalkanoate (PHA):

PHA, poly(3-hydroxybutyric acid) (P3BH), and poly(3-hydroxybutyrate-co-3-hydroxyvalerate) (PHBV: $[COCH_2CH(CH_3)O]m[COCH_2CH(C_2H_5)O]n$) are examples of some of the biopolymers which are prepared from bacteria. Diverse categories of bacterial species in the soil synthesize PHA. Early researches focused on the most basic PHA. Initiatives are taken to enhance the resin properties by varying factors such as bacterial species, genetic modifications, nutrient sources for bacteria, cultivation techniques, chemical additions, and processing methodologies. The high cost of PHA primarily is caused due to the challenges like sustenance for bacteria, the intricacy of the production method, the separation of bacteria from the biodegradable plastic, and the extraction and treatment of waste fluids. The melting point and thermal breakdown temperatures are very close. This makes the plastic melting process difficult.

Starch-based biodegradable plastics:

Starch is derived from grains, which serves as an economical source for starch production. Scientists are trying to create biodegradable polymers derived from starch by enhancing its thermoplastic properties. Starch is made up of glucose molecules. It is crystalline and remains solid without flowing or melting, even upon reaching the thermal decomposition temperature. Starch may be used as derivatives, like starch ether, esterified starch, and phosphoric acid-linked starch, and are applicable in both food and industrial contexts. Starch-based biodegradable plastic is certified as compostable and marine biodegradable, contingent upon its grade.

Cellulose-based biodegradable plastics:

The utilization of non-edible biomass, such as cellulose, to produce biodegradable polymers is an interesting and environment friendly approach. Cellulose is a major constituent of the primary component of terrestrial plants and wood. Cellulose-based polymers are less expensive due to their ease of production [21]. Researchers have made effort to degrade cellulose using acids or enzymes to produce cellulose-derived polymers like cellophane, cellulose acetate etc.

2.2 Most widely use biodegradable petroleum derived plastics:

Polybutylene adipate-co-terephthalate (PBAT):

PBAT is a thermoplastic biodegradable resin made from polyester, manufactured through the polymerization of 1,4-butanediol, terephthalic acid, and adipic acid. PBAT, is an aliphatic and aromatic petroleum-derived biodegradable copolymer. It is utilized in agricultural films and medical devices, among other applications. PBAT is flexible, ductile, has high elongation at break, and impact strength and is readily processable and are analogous to those of low-density polyethylene (LDPE). It exhibits significant biodegradability attributed to the aliphatic polyester component, while its robust mechanical strength is derived from its aromatic structure.

Polybutylene succinate (PBS):

PBS (polyester resin) is a petroleum-derived biodegradable plastic, synthesized through the condensation polymerization of succinic acid and 1,4–butanediol. It has high mechanical strength and high temperature stability among the biodegradable plastics along with non-toxicity and biodegradability in the presence of bacteria and enzymes. PBS is a home compostable plastic and is highly compatible with other biodegradable plastics such as PLA. PBS shows very good heatsealing properties. It finds applications in the agricultural films, paper cups, straws, coffee capsules, etc.

Polyvinyl alcohol (PVA):

PVA is a water-soluble petroleum-derived resin which is biodegradable in the presence of certain microbes. It demonstrates significant adhesion strength to hydrophilic materials and has very low air permeability. Saponified PVA exhibits significant crystallinity due to the existence of hydrogen bonds within its structure. PVA is typically insoluble in cold water.

Polyglycolic acid (PGA):

PGA is a biodegradable polymer made up of glycol units. The basic structure of PGA confers great strength, stability against organic solvents, and compostability. It is a nontoxic material, used in medical sutures due to its strength, biocompatibility, and biodegradability within the human body. The hydrolysis of PGA occurs rapidly. It exhibits superior mechanical qualities, enhanced resistance to organic solvents, and greater thermal stability in comparison to PLA. It is utilized in the manufacture of several categories of industrial non-woven filaments or fibers. A significant characteristic is the elevated gas barrier property of PGA.

2.3 Most widely used biomass derived nonbiodegradable plastics:

Biomass polyethylene:

Polyethylene from biomass is prepared and is been recently used to produce shopping bags in Japan. Initially, bioethanol is prepared from corn and sugarcane biomass. Then, biomass polyethylene can be prepared by polymerizing ethylene, which is obtained by dehydrating bioethanol. This PE is not biodegradable Sugarcane and corn which are the raw materials absorb greenhouse gases during their growth. The total emission of CO_2 can be considered as zero (thus carbon neutral) whenever this PE will be disposed of.

Polyamide 11:

Biomass-derived polyamide 11 is an aliphatic engineering obtained through the polymerization of 11-aminoundecanoic acid derived from castor oil. It has a melting point of around 185 °C. This polyamide has an extended methylene chain, enhancing its heat and chemical resistance. In comparison to nylon 6, polyamide 11 exhibits superior low-temperature flexibility and abrasion resistance. The number of amide linkages in PA11 is lesser compared to that of other polyamides, leading to a reduced ability for water absorption. PA11 offers superior dimensional stability, flexibility, gasoline resistance, processability, and mechanical properties without the detrimental effects of ambient moisture. Polyamide 11 is utilized in the manufacturing of vehicle fuel tubes, air brake tubes, food washing cages, computer connector covers, cable ties, and more.

Biomass PET: biomass polyethylene terephthalate:

Sugarcane biomass derived mono-ethylene glycol and terephthalic acid polymerize to prepare biomass PET. Terephthalic acid, typically obtained from p-xylene, is challenging to get from plant biomass. Biomass-based PET exhibits equivalent thermal resistance and mechanical characteristics to petroleum-derived PET. PET bottles are utilized for the storage of beverages.

Biomass polyurethane:

Polyurethanes are made up of polyol, diisocyanate as chain extenders along with additives. Castor oil is used to synthesize polyol as compared to other plant oils as it contains higher hydroxyl groups. Polyol may also be synthesized from sugar derivatives like D-glucose. On the other hand, isocyanates, such as hexamethylene diisocyanate and 1,6-diisocyanatohexane, are derived from unsaturated fatty acids, including oleic acid. Biomass polyurethane are used in the construction of buildings, biomedical applications, electronic products, and other industrial products as paints, foams, coating films, elastomers, adhesives, integral skins, and rigid plastics.

In spite of having so many options of biodegradable and biomass-derived plastics in hand, there are several challenges in effective implementation of these materials in daily life [22]. The challenges and opportunities of biodegrable plastics are shown beautifully in the diagram below.

3. Polymers recyclate:

Now a days with growing concern for the environment, a lot of initiatives are taken by the government and the public to maximize the recycling and reuse of resources. A lot of emphasis is given on the use of recycled polymers either in pure form or in mixture with virgin polymer.

The major challenges faced towards the usage of recycled polymers are:

1. Contamination: After the end use by the consumers the collection of the polymeric materials remain a huge challenge. In most of the cases polymer waste materials are collected by the waste pickers. Rag pickers are not technically sound but they collect the waste plastics based on their experiences and market demand. This leaves a scope of (a) incomplete collection (b) preferential collection (c) contamination of dirt, other polymers and composite materials. We are yet to plan out a circular economy where the collection of the plastic materials will be done through a cyclic route with some financial benefit to the customers for cooperating in the collection process and helping in the sorting of the waste at source. A bright example is the way glass milk bottles are collected and recycled everyday with the return of the old bottles. Any type of incentive always works as an encouragement for the customers to actively play a role in waste collection.
2. Degradation of polymers: Polymers when are recycled, the plastics undergo heat cycles which causes reduction in molecular weight of the polymer. This also results in lowering of the properties of the plastics. This limits the chances of using the recycled plastic for the same application.

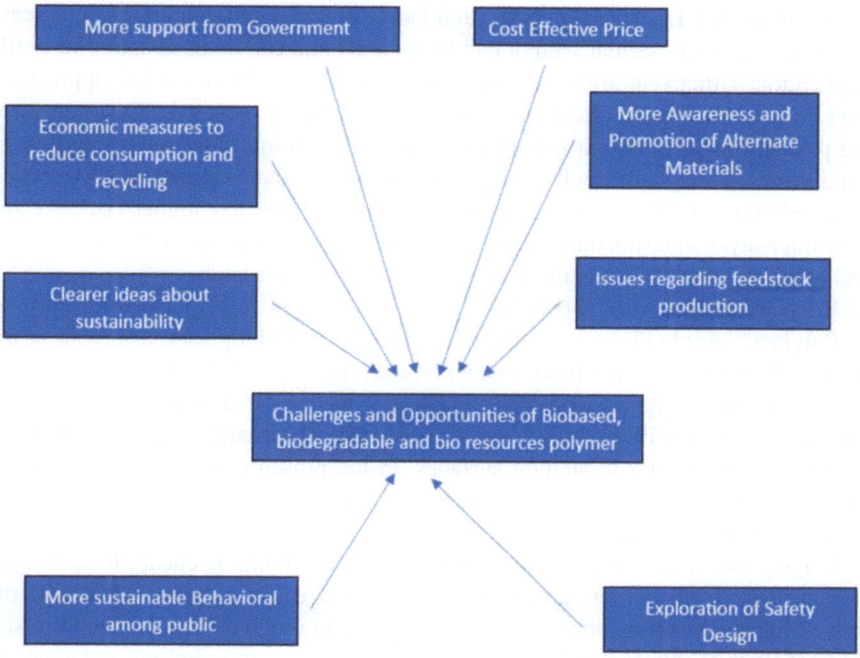

Fig. 2 Challenges and opportunities of biodegradable polymers

3. Inconsistent Infrastructure and regulations: There is a lack of any standards for recycling and tracing the recycled materials. As mentioned earlier, with every cycle of mechanical recycling, a plastic loses its properties. Since there is no tracing of the recycled plastics, there is no way to know the number of heat cycles it has gone through. Thus standardization of the recycled plastics become a challenge.
4. Compatibility issues: The recycled plastics on application of heat cycles lose their original structure. This leads to compatibility issues.
5. Public acceptance: In general public concept of recycled plastic is not good. People assume recycled material as non-hygenic and dirty. There are restrictions on the usage of recycled material in food and pharma sector.

Industries and scientific community is working towards implementing laws so that more and more recycled materials can be used in various applications.

Some of the recycled polymers which are used is various applications are discussed below:

Recycled Polyethylene terephthalate (r-PET)

PET is one of the most widely used thermoplastic polymers from the polyester family being used in our daily life. Uses of PET include fibers for textiles, biomedical applications such as scaffolds or hernia-repair mesh, composites and FDM 3D printing and packaging. The Iv of PET on recycling reduces significantly. After recycling the rPET is taken for SSP (Solid state polymerization) further to increase the Iv so that its molecular weight can be retained and can be used again for bottle application. Otherwise after recycling the rPETs received from bottle application may be used for film applications. Like other polymers, recycled PET often faces the problem of inconsistent material properties due to multiple sources and usage conditions, leading to difficult print quality and mechanical performance. Moreover, recycled PET have higher moisture content, causing issues like higher viscosity and poor inter-layer adhesion during printing. The environment demands the use of more and more recycled PET making it an important area of research.

Recycled Polylactic acid (rPLA)

Polylactic Acid (PLA) is an aliphatic biodegradable bioplastic available in the market. Post recycling the thermo-mechanical properties of the rPLA significantly change and show higher degree of crystallinity and elevated melting temperatures compared to virgin PLA. PLA is used in biomedical research, starting from bone fixation to healing the scratching surfaces. In the presence of additives and dyes, generally rPLA exhibits higher Tg values than virgin PLA.

Recycled Polystyrene (rPS)

Polystyrene (PS) is a transparent commodity plastic that is cheap. It finds wide applications in insulations and packaging. Recycled PS is quite brittle and the processing and handling is very difficult compared to virgin polymers. The immiscibility of PS with different polymer types affects the phase morphology and mechanical strength. Block copolymers like SEBS and SEBS-MA are used as compatibilizers to improve phase compatibility and interfacial adhesion of rPS, which helps in achieving better mechanical properties and printability.

Recycled Polyethylene (rPE)

Polyethylene (PE) are made of long chains of ethylene monomers and are widely commercialized for different applications, like bottles, containers, or plastic bags.

Recycled Low-density polyethylene (rLDPE)

Low-density polyethylene (LDPE) is a type of plastic which is mostly found in waste streams such as recycled garbage bags or pallet covers. The tensile strength and modulus for the rLDPE drops significantly compared to virgin LDPE.

Recycled High-density polyethylene (rHDPE)

High-density polyethylene (HDPE) is among the most utilized thermoplastics owing to its accessibility, affordability, and exceptional characteristics. HDPE has been utilized for several applications, including toys, bottles, pipes, and cable insulation. Researchers have examined recycled HDPE for 3D printing applications and documented its diverse mechanical properties. While the characteristics of recycled polymer filaments were analogous to those of commercial filaments, these investigations illustrate the feasibility of employing recycled polymers as substitutes for commercial polymer filaments.

These recycled polymers are used for certain applications and their rheological behavior needs to be understood before processing. The processing of different types of polymers varies depending on the application as well as the grades of the polymers used. In the second volume of this series, a thorough discussion of the rheological studies and processing of the polymers are discussed.

Characterization and recyclability of polymers are other important aspects which needs attention.

Role of Artificial Intelligence in Polymer Science and Technology:

Artificial Intelligence has brought a paradigm shift in every sphere of life. In today's world AI robots are found to be an efficient and cost-effective alternative to manpower. Starting from telecom business to production and research is getting more and more influenced by AI. Like all other fields polymer research, education and industry is also getting affected by AI.

AI primarily works based on the patterns like structure property relation, molecular weight, folding, primary and secondary structure and so on and helps humans to predict a certain situation/status based on the data already fed. This can be used in predictive analogy. Polymer behavior and properties are largely dependent on its molecular features which may include nature of monomer/s, percentage of comonomers, nature of arrangements, chain length and molecular weight, molecular weight distribution, secondary structures, and synthesis mechanism as well as conditions. Many times, it becomes difficult for the industry or the researchers to carry out every reaction in the lab. With the predictive behavior the AI will enable the scientist to find out the exact features of the polymer required to exhibit the expected properties. Moreover, control of the necessary conditions of the synthesis can be achieved efficiently using artificial intelligence in the reaction set up.

Similarly, the polymer processing can be maintained as well as optimized by AI. The process parameters can be easily maintained and optimized using AI. This will improve the quality of the product, will reduce resource wastage both in terms of material as well as energy making the process greener and sustainable. Usage of AI can predictz the equipment failures reducing downtime and maintenance cost.

Over all, AI improves efficiency, quality, and consistency in production processes by the application of

- Neural networks for polymerization
- Reinforcement learning for optimizing continuous processes.

During the processing of polymer AI will also ensure consistent and optimized out of extrusion, mixing and molding process. It will help in real-time process monitoring, process simulation, and predicting defects by AI-driven control systems and digital twin for extrusion lines.

As mentioned earlier, new product design requires thorough understanding of the morphology of a material, its inherent properties and structure property relations. AI can predict the correct composition, design, and formulation of a new product by

1. Material property prediction
2. Morphology and Topology optimization and
3. Generative designs.

Neural Network models and AI-based CAD tools may be used for the same.

After the end of cycle when the polymers are disposed then rises the major environmental concern which is yet to be solved. Industries are facing a lot of challenges in meeting the government policies. The major problem with plastic waste arises when it comes to sorting. Sorting of a mixed plastic waste requires the intervention from AI. AI can be applied efficiently in fast and cost-effective sorting of plastic waste and can also help humans to find the correct application based on its properties. AI is also useful in life cycle analysis of a product. It can be used for smart sorting using AI robots, recycling process optimization, and life cycle analysis.

Overall the benefits of using AI are as follows:

1. Reducing energy consumption
2. Reducing trial time
3. Optimized and faster process
4. Better product quality
5. Faster Innovation Cycle and
6. Enhanced Sustainability

The role of AI in polymer industries is only going be higher day by day. AI is predicted to be more useful in the following ways:

1. Self-Optimization processes.
2. AI driven innovation of polymers which will help in developing next generation polymers in bioplastics, electronics, and aerospace applications reducing the R&D cycle.
3. Development of Intelligent Polymers like self-healing polymers, smart materials, shape changing materials, and stimuli responsive materials helping in new breakthrough in biomedical applications, packaging applications, defence applications and many more.
4. The most important use of AI will be effective sorting and efficient degradation of materials helping the society to combat the sustainability-related issues.
5. Replacement of physical testing and simulation with AI-powered tools reducing material wastage and cost.
6. AI may help in creating unique polymer products depending on individual customer needs helping to achieve higher customer satisfaction.
7. AI can help us to solve the complicated problem of transparency and traceability of polymeric products by tracking the origin, composition, and environmental footprint of an end product.

Inspite of having so many advantages, there are challenges to be overcome in the near future, like:

1. Availability of reliable data and quality to train the AI
2. Availability of trained manpower to use AI
3. AI transparency and intellectual property problem
4. More usage of AI leading to reduced manpower requirement

We can look forward to have more application in AI in research and development in (a) polymer ageing and lifetime management, (b) in understanding and predicting polymer behavior, (c) in multiscale modelling, and (d) in AI based autonomous labs. Artificial intelligence can proved to be a boon for mankind if it is used with utmost reliability and humility.

The Future of Polymers and Environmental Impact

The field of polymer science is constantly evolving, with new advancements and emerging technologies pushing the boundaries of what is possible. Researchers are continuously exploring new materials, synthesis methods, and applications for polymers.

One area of focus is the development of smart polymers, also known as stimuli-responsive or intelligent polymers. These materials can change their properties in response to external stimuli such as temperature, pH, light, or electric fields. Smart polymers have the potential to revolutionize drug delivery systems, sensors, actuators, and other applications where controlled release or responsiveness is desired.

Another emerging technology in polymer science is additive manufacturing, also known as 3D printing. This technology allows for the direct fabrication of complex 3D objects layer by layer using computer-aided design (CAD) models. 3D printing has the potential to revolutionize manufacturing processes by enabling rapid prototyping, customization, and on-demand production.

Researchers are also exploring the use of nanotechnology in polymer science. Nanocomposites, which are materials composed of a polymer matrix reinforced with nanoparticles, offer enhanced mechanical, thermal, and electrical properties. Nanoparticles can also be used to modify the surface properties of polymers, making them more hydrophobic, antimicrobial, or conductive.

While polymers have revolutionized various industries, they also pose significant environmental challenges. The production, use, and disposal of polymers contribute to pollution, resource depletion, and waste management issues. One of the main challenges associated with polymer waste is its persistence in the environment. Many polymers are non-biodegradable and can take hundreds of years to decompose. This leads to the accumulation of plastic waste in landfills, oceans, and other natural habitats. Plastic pollution has become a global issue, with devastating impacts on marine life and ecosystems. Animals can become entangled in plastic debris or mistake it for food, leading to injury or death. Microplastics, which are tiny particles of plastic less than 5mm in size, have also been found in the air, water, and even in our food chain.

To address these environmental challenges, various solutions are being explored. One approach is the development of biodegradable polymers that can break down into harmless by products under specific conditions. Biodegradable polymers can help reduce the accumulation of plastic waste and minimize the environmental impact.

Another solution is the promotion of recycling and circular economy principles. By implementing effective recycling systems and encouraging the use of recycled polymers, we can reduce the demand for virgin materials and conserve resources. Additionally, innovative technologies such as chemical recycling and biorefineries are being developed to convert polymer waste into valuable products or feedstocks for other industries.

Polymers are truly fascinating materials that have revolutionized various industries and become an integral part of our everyday lives. Their unique properties, versatility, and ease of processing make them ideal for a wide range of applications, from consumer goods to high-tech industries.

The future of polymer science and technology holds great promise, with advancements in areas such as smart polymers, additive manufacturing, and nanotechnology pushing the boundaries of what is possible. However, it is important to address the environmental challenges associated with polymer waste and promote sustainable practices in the production, use, and disposal of polymers. As we continue to explore the creative and innovative side of polymer design, we can expect to see even more exciting applications and advancements in the field. Polymers will continue to shape our world and play a crucial role in solving some of the greatest challenges we face as a society.

References

https://www.britannica.com/biography/John-Wesley-Hyatt

https://www.britannica.com/biography/Henri-Braconnot

https://www.britannica.com/biography/Charles-Goodyear

https://www.britannica.com/biography/Leo-Baekeland

https://www.nobelprize.org/prizes/chemistry/1953/staudinger/facts/

https://www.acs.org/education/whatischemistry/landmarks/carotherspolymers.html

https://en.wikipedia.org/wiki/Phillips_Petroleum_Company

Luckachan, G.E. and Pillai, C. (2011) Biodegradable Polymers—A Review on Recent

Kuhn CD, et al. (2007) Functional architecture of RNA polymerase I. Cell 131(7):126072

https://pubmed.ncbi.nlm.nih.gov/25036628/ 11. https://iopscience.iop.org/article/10.1088/0957-4484/18/10/105104

Fang, L., Leng, Y. and Gao, P. 2005. Processing of hydroxyapatite-reinforced ultrahigh molecular weight polyethylene for biomedical applications. Biomaterials, 26(17): 34713478.

Zhang, Q., Rastogi, S., Chen, D., Lippits, D. and Lemstra, P.J. 2006. Low percolation threshold in single-walled carbon nanotube/high-density polyethylene composites prepared by Mel processing technique. Carbon, 44(4): 778–785.

Dey, K.T. and Tripathi, M. 2010. Thermal properties of silicon powder filled High-density Polyethylene composites. Thermochim. Acta., 502(1–2): 35–42.

Manu, M.T., Soni, S., Murthy, K.R.V. and Sebastian, T.M. 2013. Ba(Zn1/3Ta2/3)O3 ceramics reinforce high-density polyethylene for microwave applications J. Mater. Sci.

Pasch, H. and Eselem, B.P.S. 2018. Branching and molar mass analysis of low-density polyethylene using the multiple preparative fractionation concept. Polym. Chem., 9:

Della, T.C., Bergami, E., Salvati, A., Faleric, C., Cirino, F., Dawson, K.A. and Corsi, I. 2014. Accumulation and embryotoxicity of polystyrene nanoplastic at a bearly stage of development of sea urchin embryos Paracentrotus lividus. Environ. Sci. Technol., 48(20): 12302–12311.

Vol. 22, No. 3, 2023 • Nature Environment and Pollution Technology https://www.chemicalsupport.co.uk/wp-content/uploads/2014/04/DUPONT-PTFE-SPECSHEET.pdf

Environ. Sci. Technol. 2013, 47, 12, 6646–6655

Ryohei Mori, Replacing all petroleum-based chemical products with natural biomass-based chemical products: a tutorial review, RSC Sustainability, 2023,1.

TaofeeqD. Moshood, Gusman Nawanir, Fatimah Mahmud, Fazeeda Mohamad, Mohd Hanafiah Ahmad, Airin Abdul Ghani; Biodegradable plastic applications towards sustainability: A recent innovations in the green product, Cleaner Engineering and Technology, Volume 6, February 2022, 100404.

Laboratory Safety Rules

Like a two-sided sword, science and scientific knowledge can be used both as a bane and a boon in the human prosperity. Scientific experiments are useful techniques to be learnt by students to explore new things in their scientific endeavour but with caution. Without proper safety measures any reaction, process and goal may turn into a nightmare and disaster.

The topic lab safety not only refers to the hazards that might happen with the performer but it also covers broadly the care that needs to be taken into consideration while performing and experiment as well as lab maintenance. The topic of lab safety can be divided into the following subtopics:

1. Maintenance of instruments and maximise their efficiency
2. Knowledge of hazards related to chemicals and first aid measures to be taken if there is any accident
3. Personal safety

When all these three aspects are taken care of properly, the lab will be function at its maximum potential.

The basic safety steps to be followed may be summarized as follows:

- Comprehend all safety precautions before to operating any equipment.
- Remain alert for hazardous equipment conditions, processes, or actions, and report them immediately for rectification.
- Explicitly designate all storage places and maintain chemicals in appropriately labelled containers.
- Document the date on chemical containers at acquisition and upon opening.
- Monitor the expiration dates of substances.
- Adhere to the designated storage requirements for chemicals.
- Acquire the essential safety protocols for managing the subsequent categories of hazardous materials: (A) Flammable (B) Corrosive (C) Toxic (D) Carcinogenic (E) Compressed Gases (F) Poisonous.
- Store chemicals in accordance with their compatibility classifications.
- Isolate halogenated waste solvents from non-halogenated solvents.

- Isolate heavy metals and solid waste for secure disposal.
- Exhibit cautionary indicators for certain dangers, including flammable substances, and denote zones where open flames are forbidden.
- When combining solutions, introduce concentrated solutions into less concentrated ones to avert strong reactions (e.g., add acid to water, not water to acid).
- Refrain from distracting your colleagues.
- Utilize equipment exclusively for its designated function.
- Stabilize and arrange the apparatus utilized for chemical reactions to facilitate operation without displacement till the reaction is entirely concluded.

Individual Security

- Utilize removed wet benches exclusively for chemical operations. Consistently wear protective glasses or goggles when in the laboratory.
- Always don a laboratory coat or apron in the laboratory.
- Suitable gloves must be utilized as required.
- Suitable footwear must be worn in the laboratory.
- Utilize a respiratory mask as necessary.
- A solitary student should not be permitted to operate in the synthesis laboratory at any time.

Personal Hygiene

- Sanitize hands prior to exiting the laboratory.
- Never use oral suction on any pipette in the laboratory.
- Consumption of food or beverages is prohibited in laboratories or locations where chemicals are utilized or kept.
- Food must not be stored in a laboratory refrigerator. Do not consume food or beverages from laboratory glassware.
- Ensure that exposed skin is adequately protected in the laboratory.

Fire Prevention

- Be cognizant of ignition sources in the laboratory and service areas (open fires, heat, electrical apparatus).
- Acquire chemicals in amounts intended for imminent use.
- Consistently store flammable liquids in designated cabinets.
- Refrain from storing incompatible reagents in proximity (e.g., acids beside organic solvents).
- Avoid the prolonged storage of ethers or analogous substances due to the risk of explosive peroxides may constitute.
- Label chemicals with the date of receipt and opening.
- Ensure that all electrical cords are in proper condition and that all electrical outlets are grounded. Stay clear of the vicinity of a fire or incident if you are unable to assist.

- Acquaint yourself with the location and status of fire extinguishers. Compromised seals indicate combustion. The extinguisher has been utilized and requires recharging.
- Refrain from use fire extinguishers unless you possess the requisite training and confidence.

Domestic maintenance

- Mitigate safety risks by ensuring the laboratory workspaces are well-organized.
- Ensure unobstructed access to the laboratory exit.
- Maintain bench tops, extracted wet benches, floors, and aisles free of superfluous materials.

Sanitize bench tops and other laboratory surfaces following each utilization. All equipment must be examined prior to utilization.

- If experiments require unattended observation, affix a note adjacent to the experimental device, Specify the compounds involved, any risks, your name, and a contact number may be contacted in the event of an emergency.
- Maintain a consistently dry laboratory floor. Promptly address spills and inform the other laboratory employees of any sliding hazards.
- Maintenance work on laboratory equipment should be conducted exclusively by authorized staff. Sink traps should be routinely cleansed with water to avert the discharge of chemicals, Odors in the laboratory. All pressurized gas cylinders must be firmly secured to a rack using chains or clamps.
- Transport vacant cylinders to the designated empty cylinder bay for retrieval emergency Protocols.
- Kindly familiarize yourself with the placement, functionality, and constraints of the following safety devices: Eye Wash Station, First Aid Kit, Fire Alarm, Fire Extinguisher.

Promptly address all minor spills. If a spill is substantial and is anticipated to present a danger to alert others in the laboratory, cease the activity or equipment if feasible, and request assistance.

- In the event of a spill involving volatile, flammable, or hazardous substances, extinguish flames and deactivate spark-generating equipment. Immediately evacuate and contact one of the aforementioned staff members for assistance in managing the spill.
- In the occurrence of a fire or explosion, summon assistance.
- Ensure unobstructed access to all safety equipment at all times.

Personal Protective Equipment (PPE)

The safeguarding of workers' health and safety in the workplace is a legal obligation mandated by the Health and Safety Executive Directive. Acquaint yourself with the safety protocols and comply with them.

- Ocular Safeguarding: In a laboratory setting, the occurrence of splashing chemicals or airborne items is always a possibility. Eye protection, including safety glasses and goggles, must be worn at all times in the laboratory.
- Laboratory Coat: The laboratory coat is intended to safeguard clothing and skin from potential chemical spills or splashes. Always don a laboratory coat in the chemical laboratory.

It is essential to maintain the protective apparel in optimal condition. Soiled and impaired garments ought to be laundered and mended or substituted. Contaminated protective attire constitutes a hazard in its own right.

- Protection for Hands: Always don protective gloves in the laboratory, particularly when manipulating chemicals. Please select a glove type based on its chemical resistance, as some are not immune to chemical contact concerning the substance being processed and specific risks associated. Prior to utilization, verify that the gloves are in satisfactory shape and devoid of holes, punctures, and tears.

The glove manufacturer/supplier and the related Material Safety Data Sheets (MSDS) are reliable sources for specific glove selection information.

Here are several recommendations for glove selection:

- PVC provides protection against mild corrosives and irritants.
- Latex offers minimal defense against irritants and restricted protection against pathogens.
- Natural rubber provides protection against moderate corrosive substances and electrical shock.
- Neoprene rubber is suitable for use with solvents, oils, and mild corrosive substances.
- Cotton absorbs moisture, maintains cleanliness, and offers minimal fire retardant characteristics.
- Thermal gloves must be utilized while manipulating small heated objects. To remove gloves, begin at the wrist and take them off towards the fingers, inverting them in the process. Refrain from touching the gloves' working surface area while removing them. Immediately wash hands after removing protective gloves.
- Footwear Safeguarding

Foot protection is intended to avert injuries caused by corrosive substances, large items, electrical hazards, and to provide traction on slippery surfaces.

Kindly wear durable footwear that fully encases the foot. These will offer optimal protection. Refrain from wearing shoes that reveal the feet in any manner.

Contents

Abbreviations

ABS	Acrylonitrile Butadiene Styrene
ATRF	Atom Transfer Radical Polymerization
FTIR	Fourier Transform Infrared
HDPE	High-Density Polyethylene
HIPS	High Impact Polystyrene
LDPE	Low-Density Polyethylene
LLDPE	Linear Low-Density Polyethylene
LVER	Linear Viscoelastic Region
MF	Melamine Formaldehyde Resin
NR	Natural Rubber
ODR	Oscillating Disc Rheometer
PA	Polyamide (commonly known as Nylon)
PAN	Polyacrylonitrile
PC	Polycarbonate
PE	Polyethylene
PEEK	Polyetheretherketone
PEG	Polyethylene Glycol
PET	Polyethylene Terephthalate
PF	Phenol Formalehyde Resin
PLA	Polylactic Acid
PMMA	Polymethyl Methacrylate
PP	Polypropylene
PPE	Personal Protective Equipment
Ppy	Polypyrrole
PS	Polystyrene
PTFE	Polytetrafluoroethylene
PTh	Polythiophene
PU	Polyurethane
PVA	Polyvinyl Alcohol
PVC	Polyvinyl Chloride
PVDF	Polyvinylidene Fluoride

SAN	Styrene Acrylonitrile
SBR	Styrene-Butadiene Rubber
SFF	Solid Free Form
TGA	Thermo Gravimetric Analysis

Chapter 1
Identification and Molecular Characterization of Polymer

Abstract Polymeric materials are made up of a number of monomer units which combine with each other chemically. The chemical structure of the polymer determines the physical and chemical properties of the polymers. Some of the polymers are soft in touch while others are hard. Although most of the polymers are white in color but depending on the absorption of the functional groups the color intensities may vary. Polymers vary not only in their feel, touch or color but they also vary in their chemical and physical behavior. The solubility parameter of the polymer must match with that of the solvent. This can be used as a tool to identify the polymer. Among other methods the study of the burning behavior of the polymer granules is very important method of identification. Among the spectroscopic methods, FTIR is useful and fast method of identifying a polymer. The methods, theory and observations of all the identification methods are discussed in this chapter.

Keywords Identification · Flame test · Density · IR · Solubility

Human civilization started with the stone age followed by the metal age, glasses, and papers. But all these materials had some challenges associated with them. Mankind was looking for a material which can be as strong as steel, as flexible as paper, stable against all environmental conditions and can fulfil all the required properties to replace these materials. In this endeavor plastic came into the picture which has a wide variety in structure, properties, and applications.

Polymer is a macromolecular structure made by the association of monomers. The small molecular weight molecules with two or more functionality can be used as the monomer units of polymers. By virtue of having large molecular weight, polymers are solid in state and has significant characteristics like, high tensile strength, low solubility, flexibility, stretchability, and many more. The most important properties of these materials are its longevity and processability. But the versatility of polymer arises from the variation in its structure, orientation, isomerism, branching, comonomer, molecular weight, heteroatom present in it, branching, crystallinity and many more.

S. D. Sadhu et al., *Source Book of Experimental Techniques and Analysis of Polymeric Materials*, https://doi.org/10.1007/978-981-92-2088-5_1

Fig. 1.1 **a** Homopolymer, **b** alternate copolymer, **c** random copolymer, **d** block copolymer, **e** graft copolymer

Depending on the nature of monomers used, polymers can be broadly categorized as homopolymers and heteropolymers/copolymers. Homopolymers are the polymers which are made up of only one type of monomer. For example, polyethylene, polypropylene, polystyrene, polymethylmethacrylate, PVC, etc.

Heteropolymers or copolymers are the class of polymers which are made up of more than one type of polymers. They may be made up of two or more than two types of monomers. For example, Nylon 66, PETE, Polycarbonate, etc. Depending on the nature of the monomers and depending on the pattern of their orientation, the polymer will exhibit different types of properties. When the different types of monomers are randomly oriented then it results in the formation of random polymers. Sometimes the different types of monomers are oriented in an alternate patter forming alternate copolymers. The monomers of same type often form blocks and these blocks of different monomers get linked with each other which are called block copolymers (Fig. 1.1).

The chemical structure of the polymer is the key to the properties that will be exhibited by the polymers. By tailoring the reaction conditions of the polymer modification, the structure of a polymer will vary. This is a very important area in polymer research and studies. Polymers can be synthesized via several different techniques like bulk polymerization technique, solution polymerization technique, emulsion polymerization technique, interfacial polymerization technique, etc.

All these polymers prepared by various synthesis techniques differ from each other in terms of their molecular weight, branching, composition and chemical properties. These polymers can be differentiated and identified based on their physical properties like, feel, color, transparency, etc.

1. Effect of Branching

Polymers having the same chemical composition but different orientation leads to the formation of different levels of crystallinity and density. With increasing branching the polymers become less packed. This results in better solubility, transparency and lower T_g.

Crystallinity: Branched polymers cannot align as closely as linear chains. This results in lower crystallinity and a decreased melting point.

Lower density: Less packed, branched polymers tend to have a lower density as compared to their linear grades.

Lower tensile strength: The reduced crystallinity and lower intermolecular forces arising from lower packing between branched chains results into lower tensile strength.

Lower viscosity: Higher branching decreases the melt viscosity of a polymer due to disruption of chain entanglement. This makes the polymer flow easy.

Impact on processing: Highly branched polymers can be easy to process due to lower melt viscosity. This are beneficial for applications like film blowing or injection molding.

Branch length: Short branches on polymer backbone impacts on polymer properties more than that of long branches as shorter branches reduces the polymer chain packing more significantly.

Branching density: The degree or extent of branching per polymer chain directly impacts the overall polymer properties.

2. Secondary Force of Interaction

It is another important aspect which effects the final properties of polymers. These include van der Waals forces (London dispersion forces), dipole–dipole interactions, and hydrogen bonds, which vary in strength depending on the chemical structure of the polymer chain.

Impact on flexibility: The Stronger the secondary forces are between the polymer chains lesser is the chain mobility which results in a stiffer and less flexible polymer.

Effect on melting point: Any polymers with strong interaction of secondary forces between chains have higher melting points because more energy is needed to overcome these intermolecular interactions and separate the chains.

Solubility influence: The polarity of the polymer molecules due to the presence of functional groups, affects its solubility in a given solvent based on the strength of secondary forces of interaction with the solvent molecules.

3. Crystallinity

Crystallinity is another important factor that impacts the macro-properties of the polymers:

Effect on Mechanical Properties

Increased stiffness and strength: The crystalline regions present in a polymer matrix act as reinforcing points, enhancing the overall stiffness and tensile strength of the polymer.

Reduced ductility and toughness: Higher crystallinity of the polymer matrix leads to brittleness which means that the polymer is more prone to fracture under stress.

4. Thermal Properties

Higher melting point: Crystalline regions having higher melting point as compared to that of amorphous regions, results in higher overall melting point for the polymer.

Improved thermal stability: Polymer having higher crystallinity enhances the resistance to thermal degradation.

5. Optical Properties

Decreased transparency: Crystalline regions present in a polymer matrix scatters the light more effectively. This makes the polymer appear opaque or cloudy.

6. Solubility and Permeability

Reduced solubility: Crystalline regions are not easily solubilized by solvents as the chains are tightly packed.

Lower permeability: The ordered structure of the polymer chains/crystals hinders the diffusion of small gas or solvent molecules through the polymer.

Presence of hetero atoms in the polymer structure influences very important properties of a polymer.

7. Polarity: Introduction of heteroatoms, like oxygen, nitrogen, chlorine or sulfur, creates a polar site in the polymer structure, resulting in a stronger intermolecular interactions like dipole–dipole forces or hydrogen bonds, which increases the polarity of the polymer and affects its solubility in different solvents.

Glass transition temperature (T_g): Higher polarity of the polymer chain due to heteroatoms can lead to higher T_g. This means the polymer is more rigid at lower temperatures.

8. Mechanical Properties

(a) Increased strength: Due to the presence of heteroatoms, the chains grow polarity. This enhances the mechanical strength of the polymer due to stronger interchain interactions.
(b) Reduced flexibility: Depending on the nature of the heteroatom and its placement, polymer chains may have reduced flexibility making it more brittle. Solubility: The presence of heteroatoms may influence the polymer's solubility, making it highly soluble in polar solvents due to increased polarity.

9. Isomerism

Polymers may have variety in structures due to variation in their configuration (geometrical isomers), chain conformations (stereo isomers), and side group. These factors play very important role in deciding the final properties of the polymers.

Chain packing and crystallinity: Isomeric regularity in polymeric structure helps in packing of the chains in a more efficient manner, resulting in better crystallinity and enhanced mechanical properties like strength and stiffness. compared to irregular arrangements of chains with the presence of different isomers. The single bonds have the freedom to rotate around its own axis to arrange in such a way which will lead to better stability and better organization.

Cis–trans isomerism: In polymers having double bonds, the arrangement of side groups as cis or trans can greatly influence its properties. For example, natural rubber (cis-polyisoprene) is flexible due to its less regular chain packing, while gutta-percha (trans-polyisoprene) is rigid and crystalline.

Side chain isomerism: The position of a side group on a polymer chain determine its properties. Isomers having groups at different positions may have different melting points, solubility and glass transition temperatures.

The above discussed factors with variation in chemical structures results in a large number of plastics with a wide range of properties. Based on the chemical and physical properties the polymers can be identified. In this chapter a number of identification techniques are discussed in detail along with its procedure and theory, e.g.

1. Preliminary examination
2. Elemental analysis (Lassaigne's Test)
3. Solubility test
4. IR analysis (KBr disc)
5. Flame Test/Melting Test
6. Specific gravity determination.

There are other advanced characterization techniques which require expensive instruments. They will be discussed in other volumes of this series.

Experiment No. 1.1: Identification of Polymers

Aim: To Identify a given polymer

Theory

Polymers are chain of monomers attached to each other through chemical bonds in a particular sequence. The nature of the monomers, their sequence and the functional groups present in the structure defines the chemical, physical, mechanical, thermal, and optical properties. Polymers may be characterized and identified based on their chemical and physical properties using various techniques like, preliminary identification tests, Preliminary examination, elemental analysis (Lassaigne's Test), solubility test, IR analysis, NMR analysis, UV analysis, flame test/Melting Test and specific gravity determination methods to mention a few. Here the commodity plastics and commonly used rubbers are considered and their identification methods are discussed.

The techniques used to identify a polymer are as follows:

1. Preliminary examination
2. Solubility test
3. IR analysis (KBr disc)
4. Specific gravity determination.
5. Flame Test/Melting Test
6. Elemental analysis (Lassaigne's Test)

Experiment 1.1A: Preliminary Examination

A polymer may, upon initial assessment, be categorized into one of the following groups:

a. Rubber
b. Flexible thermoplastic
c. Rigid thermoplastic
d. Thermoset.

Manufacturer's Labeling Code: The composition of a growing number of plastic products is determined by the SPI (Society of Plastics Industry) recycling code, typically embossed on the product's base. The numeral is frequently encapsulated within the triangular arrows recycling emblem. The code is as follows:

1. PETE—Polyethylene Terephthalate
2. HDPE—High-Density Polyethylene
3. V—Vinyl/Polyvinyl Chloride (PVC)
4. LDPE—Low-Density Polyethylene
5. PP—Polypropylene
6. PS—Polystyrene
7. Others.

Texture: Teflon and polyethylene exhibit a waxy texture. Thermosets exhibit hardness and produce a brittle sound when impacted or dropped with a metallic object. Rubbers are pliable and occasionally elastic.

Color: The majority of polymers are offered in an extensive spectrum of colors due to the incorporation of diverse pigments. Without additions, pure amorphous polymers exhibit great optical clarity (transparency), while crystalline polymers range from translucent to opaque. Examples of translucent polymers include PET (polyethylene terephthalate), PMMA (polymethyl methacrylate), PC (polycarbonate), and PVAc (polyvinyl acetate).

Experiment No. 1.1B: Solubility

Thermoplastics dissolve in suitable solvents, while thermosets may expand but remain insoluble. Plastics may necessitate prolonged exposure to a solvent for dissolution. Agitation markedly decreases dissolution time. The subsequent table demonstrates the solubility behavior of polymers. The principle of like dissolves like applies; hydrocarbon solvents typically dissolve non-polar polymers, while polar solvents generally dissolve polymers with polar functional groups.

Plastic	Solvents			
	Gasoline	Toluene	Acetone	Cyclohexanone
ABS	SW	S	S	S
LDPE, HDPE	I	I	I	I
PA (polyamides)	I	I	I	I
PET	I	I	I	I
PMMA	I	SW	S	S
PP	I	I	I	I
PPO	I	SW	S	S
PS	S	S	S	S
PVC	I	SW	SW	S

I = insoluble SW = swells S = soluble

Solubility behavior is shown at room temperature. Heating can solubilize thermoplastics such as PE and PP.

Experiment No. 1.1C: IR Analysis

The infrared spectrum of a polymer sample, prepared as a KBr disc, will reveal prominent functional groups including carbonyl stretches (polyesters), N–H stretches (polyamides), aromatic bends (polystyrene), alkene stretches (cis-1,4-polyisoprene), O–H stretches (poly(vinyl alcohol)), C≡N stretches (polyacrylonitrile), and C-H stretches (polyethylene).

The infrared peaks typically observed in polymers are listed in Table 1.1.

Experiment No. 1.1D: Specific Gravity

The relative density of a polymer is crucial in ascertaining its identification. Polyethylenes, polypropylenes, and polyallomers (block copolymers of ethylene and propylene) exhibit buoyancy in water (specific gravity < 1.0), while nearly all other

Table 1.1 The functional groups and FTIR peaks associated with the groups

Chemical bond	Functional group	IR peak position (cm^{-1})
C–H	Alkane	3000–2850 (s)
	$-CH_3$	1450 and 1375 (m)
	$-CH_2-$	1465 (m)
	Alkene	3100–3000 (m), 1000–650 (s)
	Aromatic	3100–3050 (s), 900–690 (s)
	Alkyne (terminal)	3300 (s)
	Aldehyde	2850 and 2750 (w)
C=C	Alkene	1630 (w-m)
	Aromatic	1600–1475 (w-m)
C–C	Alkyne	2150–2100 (w-m)
C=O	Aldehyde	1740–1720 (s)
	Ketone	1725–1705 (s)
	Carboxylic acid	1730–1700 (s)
	Ester	1750–1730 (s)
	Amide	1670–1640 (s)
	Anhydride	1810 and 1760 (s)
	Acid chloride	1800 (s)
C–O	Alcohol, ether, ester, carboxylic acid, anhydride	1300–1000 (s)
O–H	Alcohol, phenol	
	Free	3650–3600 (m)
	Hydrogen-bonded	3500–3200 (m)
	Carboxylic acids	3300–2400 (m)
N–H	Primary and secondary amine and amide	3550–3060 (m-s); 1640–1550 (m-s)
C–N	Amine	1250–1000 (m-s)
C=N	Nitrile	2260–2240 (m)
N=O	Nitro ($R-NO_2$)	1550 and 1350 (s)
C–Cl	Chloride	800–600 (s)

non-cellular polymers exhibit sinking behavior. While the incorporation of fillers can alter the relative density of polymers, this approach nonetheless restricts the range of available options. A comprehensive technique is enclosed (Table 1.2).

Table 1.2 Specific gravity of some common polymers

Polymers	S.G.	S.G. (Range)	Polymers	S.G.	S.G. (Range)
Polystyrene	1.06	1.04–1.10	Polyethylene, L. D.	0.91	0.91–0.92
Poly(vinyl chloride)	1.25	1.16–1.35	Polyethylene, H. D.	0.95	0.94–0.96
Silicone	1.75		Polypropylene	0.89	
Nylon	1.15	1.13–1.20	Polycarbonate	1.20	
Polyurethane	1.18	1.11–1.25	Poly(vinyl chloride)	1.25	
PMMA	1.18				

Experiment No. 1.1E: Flame Tests/Melt Tests

Flame tests ascertain the ignition propensity, combustion rate, flame coloration, soot production, and olfactory characteristics of combustion byproducts (Tables 1.3 and 1.4).

Table 1.3 Flame test behavior of various elastomers

Nature of sample	Characteristics on burning	Characteristic smell	Fume color	Fume color
Natural rubber	Continuous burning	Distinctive smell organic matter hair	Black fumes	Material melts after burning and residue is tacky
Nitrile Rubber	Continuous burning with minor crackling	Acrylic pungent unpleasant smell	Black fumes giving black soot	Residue is tacky
Styrene butadiene rubber (SBR)	Burns continuously with minor crackling on flame	Unpleasant wax burning smell	White fumes	Not easily extinguishable. Residue non-tacky
Silicone	Inconsistent burning	Mild sweet smell (peroxide)	White fumes	Self-extinguishing immediately after removal from flame, goes to a white ash
Viton	Very difficult to ignite	Flame resistant, self	Extinguishing mild sweet smell (cinnamon) white fumes	Inconsistent burning mild sweet smell

Table 1.4 Flame test behavior of various thermoplastics

Name of the polymer	Continues to burn after removal of flame source				Burns, but extinguishes on removal of flame source				Additional information
	Color of the flame	Difficulty of burning	Drips	Odor	Color of the flame	Difficulty of burning	Drips	Odor	
PMMA	Blue, yellow tip	Hard	No (cast)	Fruity					No smoke during burning, flame may
			Yes (molded)						Spurt if rubber modified
Nylon 6	Blue with yellow tip	Hard	Yes	Burnt wool					
Nylon 66					Blue with yellow tip	Hard	Yes	Burnt wool or hair	More rigid than Nylon 6
PC				Phenol	Orange or yellow	Hard	Yes	Faint, sweet aromatic ester	Black smoke with soot in air
Polyethylene (PE)	Blue with yellow tip	Easy	Yes	Paraffin					Floats in water
Polypropylene (PP)	Blue with yellow tip	Hard	Yes	Acrid or diesel fumes					Floats in water; more difficult to scratch than polyethylene
Polystyrene (PS)	Yellow	Easy	Yes	Illuminating gas or marigold					Dense black smoke with soot in air
Polyvinyl chloride (PVC)	–	–	–	–	Yellow with green edges	Hard	No	Hydro-chloric acid	White smoke

Experiment No. 1.1F: Elemental Analysis (Lassaigne's Test)

The identification of nitrogen (e.g., polyamides), halogens (e.g., PVC or poly(vinylidene chloride)), and sulfur (vulcanized rubber and polysulfide crosslinks) can be accomplished using inorganic qualitative analysis following sodium fusion. Additives may disrupt; for instance, certain flame retardants contain chlorine or bromine.

Procedure

A. Specific Gravity Test: The specimen is weighed in air (a). The sample is then weighed when immersed in distilled water at 23 °C (b) using a sinker and wire to hold the specimen completely submerged as required.

Density and Specific Gravity are calculated.

Specific gravity = a/[(a + w)-b]
a = mass of specimen in air.
b = mass of specimen and sinker (if used) in water.
W = mass of totally immersed sinker if used and partially immersed wire.
Density, kg/m^3 = (specific gravity) × (997.6).

B. Solubility Test: A few granules of the given polymer are dipped into 15 mL of various solvents like, gasoline, toluene, acetone, cyclohexane and water with constant stirring at room temperature for 1 h. If there is no change/swelling that means that the polymer is not soluble in that solvent. If it is swelled, then the polymer may be stirred for another 1 h to get soluble completely.
C. Flame Test: A few granules of the polymer sample/a small piece of the polymer is taken in a tong and hold it on the flame. The flame color, melting behavior and extinguishing nature of the samples may be noted.
D. Lassaigne's Test: A small piece of Na metal is heated in a fusion tube with the organic compound. In doing so, Na converts all the elements present in the compound/polymer into ionic form.

- $Na + C + N \rightarrow NaCN$
- $2Na + S \rightarrow Na_2S$
- $Na + X \rightarrow NaX$ (X = Cl, Br, or I).

The formed ionic salts are extracted from the fused mass by boiling it with distilled water. This is called sodium fusion extract.

Tests can be conducted to identify these ionic compounds:

a. Test for Nitrogen

The extract is boiled with $FeSO_4$ and acidified with concentrated H_2SO_4. The appearance of Prussian blue color indicates the presence of nitrogen.

The following reactions occur,

1. $Fe^{2+} + 6CN^- \rightarrow [Fe(CN)_6]^{4-}$
2. $Fe^{2+} + H^+ \rightarrow Fe^{3+} + e^-$
3. $[Fe(CN)_6]^{4-} + 4Fe^{3+} \rightarrow Fe_4[Fe(CN)_6]\cdot H_2O$

The acid helps in the oxidation of ferrous ions to ferric ions. The formation of ferriferous cyanide indicates the presence of nitrogen.

Note: This test is not given by compounds containing N but not C atoms. For example, NH_2NH_2 does not answer this test despite having an N atom. This is because both C and N are required to form CN^- ion.

This test is not given by diazonium salts as they decompose to give nitrogen gas on heating.

b. Test for Sulphur

The extract is treated with sodium nitroprusside. The appearance of violet color indicates the presence of sulphur.

The following reaction occurs:

$$S^{2-} + [Fe(CN)_5NO]^{2-} \rightarrow [Fe(CN)_5NOS]^{4-}$$

c. Test for Halogens

The extract is acidified with HNO_3 and then treated with $AgNO_3$. A white precipitate soluble in NH_4OH indicates the presence of Cl, a yellowish precipitate sparingly soluble in NH_4OH indicates the presence of Br, and a yellow precipitate insoluble in NH_4OH indicates the presence of I.

$$AgNO_3 + NaX \rightarrow AgX \downarrow + NaNO_3$$

Observations

a. Check the feel of the material

Sample no.	Opacity	Soft/ hard	Elastic/ rigid	Colour

b. Check Solubility of the given polymer

Sample	Gasoline	Toluene	Acetone	Cyclohexanone	Water

c. Elemental Analysis

Sample	Nitrogen	Halogen	Sulphure	No hetero atom

d. Density/Specific Gravity

Sample	Lighter than water	Heavier than water

e. Polymer Identification—Flame Test

Polymer	SE or P	Speed of Burn (slow, med., fast)	Chars (C) or Melts (M)	Drips	Flame Color	Smoke Color	Sinks (□) or Floats (□) in water	TP or TS	REMARKS

SE = Self Extinguishing, P = Propagating; TP = Thermoplastic, TS = Thermoset

f. FTIR study.

Sample	Peak I	Peak II	Peak III	Peak IV	Peak V

Result: The given polymeric sample is identified as......................

Precautions

1. While checking solubility, the plastic granules should be stirred at room temperature for at least 10–15 min. If there is not at all any change in the sample then only increase the temperature slightly above room temperature and repeat the stirring process.
2. Note the color of the flame and the dripping behavior carefully. 3. Carefully handle the solvents and the flame

Experiment No. 1.2: Determination of Viscosity Average Molecular Weight of Polymers by Ubbelohde Viscometer

Aim: Determination of viscosity average molecular weight of polymers by Ubbelohde viscometer.

Apparatus	Chemicals
Ubbelohde viscometer	PVA Solution (1%) in water
Temperature-controlled water bath	
Stopwatch or timer	
Suction device	

*Similar experiments can be carried out with PET, PVC, etc. with their respective solvents Theory

The Ubbelohde viscometer is a capillary based instrument that evaluates viscosity by sucking liquid via capillary and measuring bulb. It operates on the principle that a liquid flows through a capillary tube, and the time taken for a certain volume of the liquid to pass through the tube is directly related to its viscosity. It is generally recommended for high viscosity solutions of cellulosic polymers. The Ubbelohde viscometer was developed by the German chemist Leopold Ubbelohde. This viscometer shares similarities with the Ostwald viscometer, particularly in its U-shaped glass design and basic operating principles. Both are used to measure the viscosity of liquids based on how long it takes for a liquid to flow through a capillary tube under the force of gravity. The Ubbelohde viscometer's key advantage is that the measured viscosity values are independent of the total volume of liquid in the system. This makes it more accurate for viscosity measurements, as variations in liquid volume do not affect the result. The device works by introducing a liquid sample into the reservoir, which is then drawn through a measuring bulb and capillary by suction. After that, the liquid is allowed to flow back through the calibrated marks in the measuring bulb, and the time taken for this to happen is used to determine the viscosity. This device has a third arm, open to the atmosphere, ensures that the pressure driving the liquid through the capillary is always consistent. This eliminates the dependency on the total volume of the liquid in the reservoir. The Ubbelohde viscometer's design fixes the pressure head at a constant level, which improves accuracy by making the measurements unaffected by variations in liquid volume. The Ubbelohde viscometer helps determine the intrinsic viscosity, which correlates to the molecular weight of polymers. It's used to check the viscosity of oils, ensuring quality and performance in various industries such as automotive, petrochemical, and aerospace. Useful in quality control for fluid products like syrups or gels. The viscosity average molecular weight of polymers can be evaluated using Ubbelohde viscometer by analyzing the relative viscosity and determining intrinsic viscosity using the equation $[\eta] = KM\alpha$.

ASTM D2857 is the standard practice for dilute solution viscosity of polymers which describes the method to determine the dilute solution viscosity of polymers using capillary viscometers such as the Ubbelohde viscometer. ASTM D446 is the standard specifications and operating instructions for glass capillary kinematic viscometers that specifies the dimensions, calibration, and operation of glass capillary viscometers, including Ubbelohde viscometers. ASTM D2515 is the standard test method for determination of viscosity of polymers in dilute solution using capillary viscometers which provides guidelines for measuring the viscosity of polymer solutions to calculate molecular weight (Fig. 1.2).

Procedure

1. Clean the viscometer thoroughly with an appropriate solvent to remove any contaminants.
2. Rinse it with the sample liquid (to avoid dilution with any leftover solvent).
3. Introduce the liquid sample into the viscometer through the larger reservoir tube (the sample arm). The liquid level should be sufficient to rise into the capillary when suction is applied but not too high to affect the accuracy of the measurement.
4. Ensure the liquid does not spill over into the third arm (which is open to the atmosphere).
5. Insert the viscometer into a thermostatic bath, typically set to the desired measurement temperature (e.g., 25 or 40 °C for most standard measurements).
6. Allow the viscometer and the liquid sample to equilibrate at the set temperature (usually 10–15 min). This is important because viscosity is highly temperature dependent.
7. Using a suction device, draw the liquid sample up through the capillary and into the measurement arm (the smaller side of the U-shaped tube). The liquid should rise above the upper calibration mark on the measurement bulb.
8. Remove the suction and allow the liquid to flow back down through the capillary under gravity.
9. Start the stopwatch as the liquid passes the upper calibration mark (on the measurement arm of the viscometer).
10. Stop the timer when the liquid reaches the lower calibration mark.
11. Record the flow time, which is the time taken for the liquid to travel between the two calibrated marks.
12. Repeat the measurement at least two or three times to ensure accuracy.
13. The average of these flow times is used for calculating the kinematic viscosity.
14. After completing the measurements, clean the viscometer using a suitable solvent, and let it dry to avoid contamination for future tests.

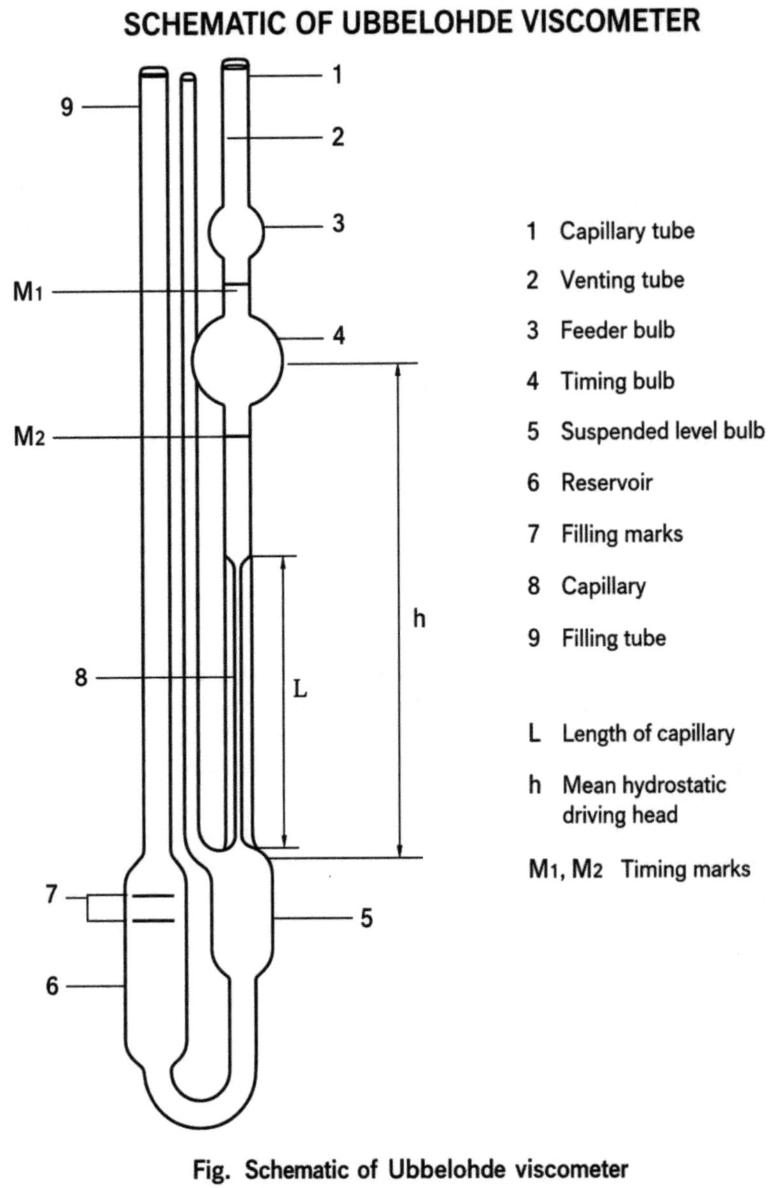

Fig. 1.2 Depiction of Ubbelohde viscometer

Table 1.5 Variation of viscosity with concentration

Sl. No.	Flow time (t_o)	Flow time of solution (t)	Concentration n (c)	$\eta_R = t/t_o$	$\eta_{sp} = \eta_R - 1$	$\eta_{sp} = \eta_{Red}$	η of R/c
1							
2							
3							
4							
5							

Observations

See Table 1.5.

I. Solvent as water

Sl. No.	Bulb 1	Bulb 2	Bulb 3
1			
2			
3			

$t_o =$ ____ (flow time)

II. Polymeric solution

Sl. No.	Bulb 1	Bulb 2	Bulb 3
1			
2			
3			

$t_1 =$ ____ (flow time)

III. Polymeric solution with 5 mL water

Sl. No.	Bulb 2
1	
2	
3	

$t_2 =$ ____ (flow time)

IV. Polymeric solution with 10 mL water

Sl. No.	Bulb 2
1	
2	
3	

$t_3 =$ ____ (flow time)

V. Polymeric solution with 15 mL water

Sl. No.	Bulb 2
1	
2	
3	

$t_4 =$ ____ (flow time)

VI. Polymeric solution with 20 mL water

Sl. No.	Bulb 2
1	
2	
3	

$t_5 =$ ____ (flow time)

Calculations

$[\eta] = KM\alpha$

$\eta =$ ____

$M =$ ____

$\alpha =$ ____

$M = [\frac{\eta}{K}]^{1/\alpha}$

Suppose 0.5% of polymeric solution is used

Then,
5 g of polymer in 100 mL of water

Therefore,
Concentration = 0.05 g/mL

We know,

$$C_1V_1 = C_2V_2$$

$$C_2 = C_1V_1 / V_2$$

$$C_2 = \_\_\_\_\_\_\_\_$$

Similarly,

Calculate C_3, C_4, and C_5 using the same formula

$$C_2V_2 = C_3V_3$$

$$C_3 = C_2V_2 / V_3$$

$$C_3 = \_\_\_\_\_\_\_\_$$

$$C_3V_3 = C_4V_4$$

$$C_4 = C_3V_3 / V_4$$

$$C_4 = \_\_\_\_\_\_\_\_$$

$$C_4V_4 = C_5V_5$$

$$C_5 = C_4V_4 / V_5$$

$$C_5 = \_\_\_\_\_\_\_\_$$

Calculating Kinematic Viscosity: The kinematic viscosity (ν) is calculated by multiplying the flow time (t) by a viscometer constant (C), which is specific to each instrument. The formula is:

$$v = C \cdot t$$

where

v = kinematic viscosity (in centistokes, c. St)

t = average flow time (in seconds)

C = viscometer constant (provided by the manufacturer).

Calculating Dynamic Viscosity

- If needed, you can calculate the dynamic viscosity (η) by multiplying the kinematic viscosity (ν) by the density (ρ) of the liquid:

$$\eta = \nu \cdot \rho$$

where

η = dynamic viscosity (in Poise)

ρ = density of the liquid (in g/cm^3).

Graph showing variation of viscosity with concentration.

Result

Molecular weight of given polymer is __________ g/mol.

Precautions

1. Since viscosity is temperature-sensitive, always ensure the liquid and viscometer are in equilibrium with the bath temperature.
2. Taking multiple readings improves the reliability of your results.
3. The viscometer should be calibrated with a standard liquid of known viscosity to ensure accuracy, especially if you're measuring viscosity values for critical applications.

References

1. https://www.boedeker.com/Technical-Resources/Technical-Library/PlasticIdentification
2. https://www.ji-horng.com/plastic-material-identification-by-burn-test

Chapter 2
Polymerization Techniques

Abstract Polymers are synthesized by adding one monomer to the other in a systematic manner. The process of adding up the monomers is of various types leading to several variations in structure, molecular weight, branching and molecular weight distribution. Among the various methods of polymerizations, following are the important ones, radical polymerization, suspension polymerization, interfacial polymerization, emulsion polymerization, solution polymerization etc. In this chapter different types of homopolymers and copolymers are synthesized by the above-mentioned methods. Both thermoplastics and elastomers synthesis techniques are discussed in this section.

Keywords Polymerization · Bulk polymerization · Solution polymerization · Emulsion polymerization · Interfacial polymerization

Polymers are a unique class of materials which are known for their versatility, unique properties, and applications. Before the starting of polymer age, mankind was primarily dependent on metals and glasses. But the challenge with these materials was their longevity and lack of ease of handing. With the discovery of polymers mankind was blessed with a class of material which meets all the benefits of metals and glasses as well as shows unique blend of properties with ease of handling and longevity.

Polymers are made up of large number of monomers arranged in a chainlike structure through chemical bonds. The chemical nature of the polymer is strictly dependent on the monomer nature, while other properties are dependent not only on the type of monomer but also on the way they are linked together, their spatial arrangement and interchain as well as intrachain interactions.

Bulk Polymerization: In bulk polymerization is done by mixing the monomer with free radical initiator and heat it. This reaction follows radical polymerization process. The increase in viscosity is an indication of the progress of the reaction. Chain transfer agents are used when the molecular weight is to be controlled. At higher viscosity the heat entrapment and the mixing of the solution become a challenge. On prolonged

S. D. Sadhu et al., *Source Book of Experimental Techniques and Analysis of Polymeric Materials*, https://doi.org/10.1007/978-981-92-2088-5_2

reaction time, the viscosity of the solution abruptly increases significantly and forms gel. The restricted diffusion of the growing polymer chain, due to the higher viscosity of the medium is the biggest challenge of this process. Lack of proper mixing and heat dissipiation is utmost important to avoid the auto-acceleration mechanism and gelation. Proper cooling arrangements should be made to handle this issue.

Advantages of Bulk Polymerization

- Simple process
- higher purity of the polymer obtained
- less number of chemicals required
- direct utility of the polymer with no isolation requirement (due to higher purity).

Emulsion Polymerization: Emulsion polymerization process is used to prepare polymer latex. In this polymerization technique the emulsified monomer is dispersed in aquous phase along with surfactants, protective colloids and some buffers (to stabilize the emulsion, like anionic surfactants or cationic surfactants or alkyl ammonium halides or nonionic surfactants). These surfactants reduce the surface tension at the interface of water monomer but due to low solubility, surfactants are sparsely soluble in aqueous medium. When the excess un-dissolved molecules form aggregates called micelle are formed an equilibrium is reached between the dissolved surfactant molecules and the aggregated micelles. The polar end of the emulsifier molecular ends align outward and the hydrocarbon ends come close to each other at the interface of the micelle.

The assembly of the hydrocarbon ends of the emulsifier molecules, make the interiors of the micelles hydrophobic, making the solubilization of monomers easy. The initiator molecules which are hydrophilic in nature are present at the surface whereas the interior is filled with the monomer. Presence of initiators at the surface of the micelle cause the polymerization to start at the surface of the micelle and proceeds inwards. More and more monomer diffuses inside the micelle from the aqueous phase as the monomer in the micelle gets consumed. This chain growth polymerization continues till another radical arrests the chain growth. With higher number of polymers formed, the chains form aggregate/fine particles. The emulsifiers stabilize these particles at the interface. The resultant emulsion obtained after polymerization is a milky white dispersion, which can be used as adhesives, emulsion paints or the polymer can be isolated from the latex by the destabilization of the emulsified polymer mass using electrolytes or by spray drying or by freezing.

Advantages

- Reaction in emulsion form helps in managing the exothermic heat of reactions and high viscosity problems due to ease of stirring of the continuous aqueous phase.
- The resultant latex may be used directly without purification (Fig. 2.1).

Solution Polymerization: This technique involves the addition of the free radical initiator in the monomer in a suitable solvent. Among important reagents, ionic co-ordination catalyst and the chain transfer reagent are needed. Viscosity control

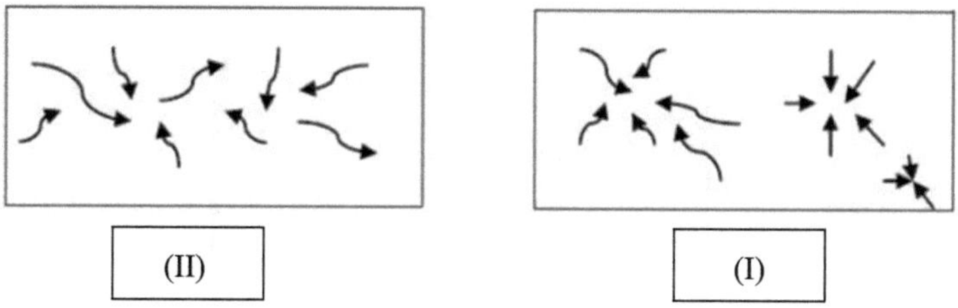

Fig. 2.1 **I** Random aligned but complete distribution at low concentration. **II** Micelles in the schematic diagram of emulsion polymerization

and prevention of heat transfer is easier in presence of an inert solvent. One of the advantages of this method is that the polymer is obtained in solution form.

Advantages

- Solvent acts as a diluent and helps in release of heat of polymerization.
- Solvent reduces viscosity, making processing easier.
- Heat control is easier than in the bulk.

Disadvantages

- Due to chain transfer reactions low molecular weights chains are produced.
- Difficult to remove solvent from final form, causing degradation of bulk properties.
- Leads to environmental pollution due to solvent release.

Suspension polymerization: In this method polymerization of hydrophobic monomers. Is done by suspending the monomer droplets in water and stabilizing this suspension through the use of stabilizers, surfactants, protective colloids, and agitation. The reaction initiators in this case are soluble in monomers. Each isolated monomer droplet functions as a distinct nucleus for bulk polymerization. The heat of the reaction and the temperature is regulated due to the continuous aqueous phase that serves as an effective heat transmission medium. The final outcome of the reaction largely depends on the ratio of monomer to water, the type and concentration of stabilizers, and the method and velocity of agitation etc. The polymerization may go up to 100% conversion, resulting in the formation of 'spherical beads or pearls'. This process is referred to as bead or pearl polymerization.

The polymer may be obtained through the filtration of the pearls or beads, followed by the elimination of surfactants and other additives via a straightforward washing with water.

Advantages

- Low viscosity due to the suspension
- Easy heat removal due to the high heat capacity of water
- Polymerization yields finely divided, stable latex sand dispersions to be used directly in coatings, paints, and adhesives (Fig. 2.2).

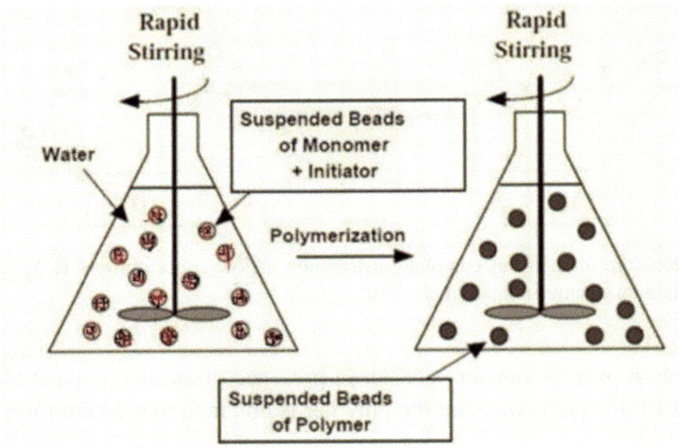

Fig. 2.2 Schematic representation of suspension polymerization

Interfacial Polymerization

From the name interfacial polymerization it can be understood that the polymerization happens at the interface between two immiscible phases, such as liquids. Typically, two monomers dissolved in two immiscible layers are brought in contact at the interface where the monomers react. As the polymer is removed from the interface more and more monomers come in contact and react with each other. This way the reaction continues to happen until one of the monomer gets exhausted. It's a special step-growth polymerization which can result into polymers with unique properties e.g. anisotropic shapes, hollow structures, and alternative surface chemistry. In interfacial polymerization, one phase contains a nucleophile reactant, such as amines or alcohols, and the other phase contains an electrophile reactant, such as acid chlorides or isocyanates. Interfacial polymerization can be used to make films, capsules, and fibers for use as separation membranes and electrode materials.

Advantages

- The reaction rate in Interfacial polymerization is very fast at a low reaction temperature. This creates heat-resistant polymers with high melting points.
- Interfacial polymerization is used to produce ultrathin layers, hollow nanospheres, and nanofibers structures.
- Interfacial polymerization is used to prepare a wide range of polymers, which includes polyamides, polyurethanes, polyurea, polyanilines, polyimides, and polycarbonates.
- This type of polymerization can be used for the preparation of defect-free, ultrathin films of metal organic frameworks, organic–inorganic hybrids, and bio-hybrids.
- Thin-film composite (TFC) membranes are prepared by Interfacial polymerization using nanofiltration (NF) and reverse osmosis (RO).

Disadvantages

- Usage of toxic solvents in interfacial polymerization.
- For interfacial polymerization continuous and intensive stirring is required.
- Conventional interfacial polymerization technique is used to produce membranes with low permeability.
- Interfacial polymerization techniques also produces membranes that are susceptible to fouling.
- In special cases interfacial polymerization have a lower degree of network formation than conventional methods.

Polymer molecular weight can be controlled by controlling the reaction path or the polymerization technique:

- The factors that can control the molecular weight distribution (MWD) in irreversible step-growth polymerization, are the extent of reaction in the first stage, the final extent of reaction, and the amount of additional monomer fed.
- In step-growth polymerization, by adjusting the stoichiometric imbalance of the monomers the molecular weight can be controlled.
- Different polymerization techniques, can be used to control the Molecular Weight Distribution of polymers. Each technique has different advantages and disadvantages. For example, solution polymerization produces lower molecular weight polymers, while gas-phase polymerization produces higher molecular weight polymers.

The Molecular Weight Distribution of polymers can be controlled by different polymerization techniques. Some common polymerization techniques are bulk, solution, suspension, emulsion, and gas-phase polymerization. Each of the above technique has its own advantages and disadvantages for controlling the MWD of polymers.

For example:

- Bulk polymerization involves polymerizing monomers without any solvent or diluent, resulting in high molecular weight and broad MWD polymers.
- Solution polymerization involves polymerizing monomers in a solvent, resulting in lower molecular weight and narrower MWD polymers.
- Suspension polymerization involves polymerizing monomers in small droplets dispersed in a liquid medium, resulting in moderate molecular weight and MWD polymers.
- Emulsion polymerization involves polymerizing monomers in micelles stabilized by surfactants in a liquid medium, resulting in very low molecular weight and very narrow MWD polymers.
- Gas-phase polymerization involves polymerizing monomers in a gaseous state, resulting in high molecular weight and narrow MWD polymers.

Atom Transfer Radical Polymerization (ATRP)

Atom Transfer Radical Polymerization technique is a new technique to polymerize through radical polymerization method in a controlled manner or using living radicals to produce polymers with tailored synthesis of molecules with specific structure, controlled molecular weight and molecular weight synthesis will be possible. This is achieved through usage of reversible deactivation process, where active polymer chains are temporarily deactivated, allowing the controlled growth and chain-end functionality.

Key Features of ATRP

1. Controlled/Living Polymerization: It is a type of reversible-deactivation radical polymerization (RDRP) process allowing control over the polymerization process, similar to living polymerization.
2. Reversible Deactivation: The hidden key to ATRP process is the dynamic equilibrium switching between active radical species and dormant species (inactive chains). Transition metal catalyst and its ligand are the key to maintain this equilibrium.
3. Controlled Molecular Weight and Distribution: The control of tailoring the molecular structure and molecular weight is done by controlling the number of active radicals. ATRP enables the synthesis of polymers with predetermined molecular weights and narrow molecular weight distributions (low polydispersity).
4. Chain-End Functionality: In ATRP allows the retention of chain ends which can be functionalized for further applications.
5. Versatile Monomer Scope: This process can be applied to a wide range of monomers which includes styrenes, acrylates, and methacrylates.
6. Block Copolymers: ATRP is a powerful tool for synthesizing block copolymers due to the retention of chain ends

Steps followed in ATRP synthesis:

1. Initiation: A metal catalyst (often copper-based) in its lower oxidation state (e.g., Cu(I)) reacts with an initiator (an alkyl halide) to generate a small number of active radical species.
2. Propagation: These active radicals react with monomers, adding them to the growing polymer chain.
3. Deactivation: The active radical species are reversibly deactivated by the catalyst, forming dormant polymer chains.
4. Dynamic Equilibrium: This process of activation and deactivation is repeated, maintaining a dynamic equilibrium between active and dormant species.
5. Chain Extension: The dormant chains can be reactivated to continue growing, allowing for the synthesis of block copolymers or other complex architectures.

ATRP Versus Conventional Radical Polymerization

Sl. No.	ATRP	Conventional radical polymerization
1	Has control over molecular weight and distribution	Lacks the control over molecular weight and distribution
2	Minimizes side reactions like termination	Side reactions are common in conventional radical polymerization
3	Conventional radical polymerization can be applied to a wider range of monomers	ATRP offers better control over the polymerization of specific monomers

In summary, ATRP is a powerful and versatile technique for synthesizing well-defined polymers with controlled molecular weights, narrow distributions, and chain-end functionality, making it suitable for a wide range of applications, including the preparation of advanced materials and block copolymers.

Synthesis of both thermoplastics and thermosets are done in lab. Thermoset preparation involves preparation and curing steps of the monomers. The methods of synthesis of polymers using ATRP are discussed in the following sections.

Experiment No. 2.1: Synthesis of Polystyrene by Bulk Polymerization Technique

Aim: To synthesis Polystyrene by Bulk Polymerization Technique and study of the experimental conditions.

Apparatus	Chemicals
Separating funnel	Styrene
Reaction test tube	Benzoyl peroxide
Oil bath	NaOH
Heating mantle	

Theory

Bulk or mass polymerization is a polymerization technique where the monomers and the initiators are mixed in a container and the reaction is initiated by heating or exposing to radiation. Normally this type of reactions is exothermic in nature. As the reaction proceeds the mixture becomes more viscous. The reaction time, concentration of initiator and nature of monomer determines the product characteristics. Overheating for prolonged time may cause gel formation. In this process no other solvents are added. During the reaction it is important to control the heat of reaction.

Styrene monomer procured from market contains inhibitors which helps in stopping the auto polymerization during storage. Prior to polymerization process, the

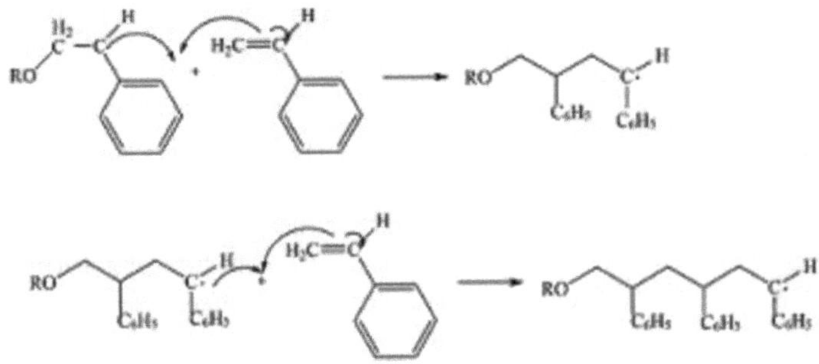

Scheme 2.1 Preparation of polystyrene polymerization

inhibitors are washed out using base followed by water. Hence, before starting the polymerization process, it's important to remove the inhibitors to smoothly carry out the polymerization of polystyrene.

To remove the inhibitors the monomers are washed with 10% NaOH solution for 3/4 times and then finally wash it with water to remove the residual base in the monomer using a separating funnel.

Reaction Scheme

See Scheme 2.1.

Procedure

1. Take 10 mL styrene in a separating funnel.
2. Add 10 mL of 10% NaOH to the separating funnel and shake well.
3. Allow the mixture to settle down and separate the two layers. The NaOH phase separates at the bottom.
4. After complete separation of the two phases, let the NaOH phase come out from the nozzle at the bottom.
5. Again repeat the same process another 3 times.
6. Now add 10 mL distilled water, and repeat the same process.
7. After washing the monomers for several times, we can collect the washed and inhibitor free monomer for polymerization.
8. Now take the monomer in a reaction test tube along with 1% benzoyl peroxide and cover it with aluminum foil.
9. Heat the reaction bath above 65 °C and continue the reaction for 20 min.
10. After the reaction is done for 20 min, the reaction mixture is poured into methanol.
11. The polystyrene is obtained as a white precipitate at the bottom.
12. The precipitate is dried and weighed.

Observation

Sl. No.	Amount of monomer (mL)	Amount of initiator (…%)	Time of reaction (min)	Percentage yield of polymer obtained (g)
1				
2				
3				

Plot a graph between

(a) Amount of monomer versus Yield of the polymer
(b) Amount of initiator versus Yield of the polymer
(c) Time of reaction versus Yield of Polymer

Result

Polystyrene is prepared by bulk polymerization method. The percentage yield of the polymer is ………….. %.

The formation of polystyrene can be confirmed using FTIR and other identification test mentioned in Chap. 1.

Precautions

1. The monomer should be completely washed to remove the inhibitors
2. The time of polymerization should be maintained properly to avoid gelling.

Experiment No. 2.2: Synthesis of Polymethylmethacrylate by Bulk Polymerization

Aim: To synthesis of Polymethylmethacrylate by Bulk Polymerization Technique and study of the experimental conditions.

Apparatus	Chemicals
Separating funnel	Methyl methacrylate
Reaction test tube	Benzoyl peroxide
Oil bath	NaOH
Heating mantle	

Theory

Bulk or mass polymerization is conducted by introducing a soluble radical initiator to pure liquid monomer. The initiator must be solubilized in the monomer. The reaction commences through heating or radiation exposure. As the reaction progresses, the mixture becomes increasingly viscous. The reaction is exothermic, resulting in the production of a diverse array of molecular masses.

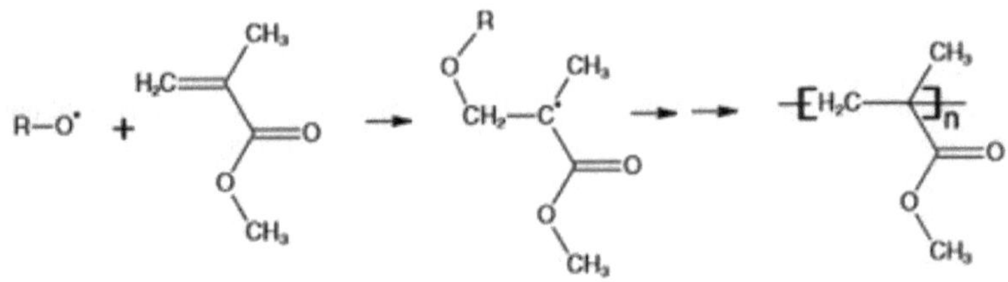

Scheme 2.2 Preparation of PMMA preparation

Bulk polymerization is conducted without any solvent or dispersion, making it the most straightforward in terms of formulation. It is utilized for the majority of step-growth polymers and other forms of chain-growth polymers. In chain-growth reactions, typically exothermic, the heat generated may render the reaction too strong and challenging to regulate without effective cooling measures.

The methyl methacrylate monomer obtained from the market contains inhibitors that prevent auto-polymerization during storage. Inhibitors must be eliminated before the polymerization process. Therefore, prior to initiating the polymerization process, it is essential to eliminate the inhibitors to facilitate the seamless polymerization of polymethylmethacrylate.

The inhibitors are eliminated by washing the monomers with a 10% NaOH solution three to four times, followed by a final rinse with water to eliminate any remaining base in the monomer using a separating funnel.

Reaction Scheme

See Scheme 2.2.

Procedure

1. Take 10 mL methylmethacrylate in a separating funnel.
2. Add 10 mL of 10% NaOH to the separating funnel and shake well.
3. Allow the mixture to settle down and separate the two layers. The NaOH phase separates at the bottom.
4. After complete separation of the two phases, let the NaOH phase come out from the nozzle at the bottom.
5. Again repeat the same process another 3 times.
6. Now add 10 mL distilled water, and repeat the same process.
7. After washing the monomers for several times, we can collect the washed and inhibitor free monomer for polymerization.
8. Now take the monomer in a reaction test tube along with 1% benzoyl peroxide and cover it with aluminum foil.
9. Heat the reaction bath above 65 °C and continue the reaction for 20 min.
10. After the reaction is done for 20 min, the reaction mixture is poured into methanol.
11. The polymethylmethacrylate is obtained as a white precipitate at the bottom.
12. The precipitate is dried and weighed.

Observation

Sl. No.	Amount of monomer (mL)	Amount of initiator (…%)	Time of reaction (min)	Percentage yield of polymer obtained (g)
1				
2				
3				

Plot a graph between

(a) Amount of monomer versus Yield of the polymer
(b) Amount of initiator versus Yield of the polymer
(c) Time of reaction versus Yield of Polymer

Result

Polymethylmethacrylate is prepared by bulk polymerization method. The percentage yield of the polymer is ………….. %.

The formation of polymethylmethacrylate can be confirmed using FTIR and other identification test mentioned in Chap. 1.

Precautions

1. The monomer should be completely washed to remove the inhibitors
2. The time of polymerization should be maintained properly to avoid gelling.

Experiment No. 2.3: Radical Copolymerization of Styrene and Methyl Methacrylate

Aim: To Radical Copolymerize of Styrene and Methyl Methacrylate

Apparatus	Chemical
10 mL pipette	Distilled water
Round bottom flask	AIBN
500 mL beakers, 400 mL beaker	Styrene, methyl methacrylate,
Medium sized stirrers	
Buchner funnel, 1 filter adapter, 1 filtering flask with hose connection	

Theory

The speeds at which various monomers incorporate into the expanding chain during the polymerization of a mixture of two or more monomers dictate the composition and, consequently, the features of the resultant copolymer. The sequence and proportion of monomer addition are dictated by their relative reactivities during the chain-growth phase, which are, in turn, affected by the characteristics of the chain's terminus, contingent upon the previously included monomer. The options include random and systematic addition, as well as block construction.

With two monomers available, four potential propagation processes exist, assuming that growth is only determined by the characteristics of the terminal end of the developing chain and the monomer.

$$M1\cdot + M1 \blacktriangleleft \quad K_{11} \tag{2.1a}$$

$$M1\cdot + M2 \blacktriangleleft \quad K_{12} \tag{2.1b}$$

$$M2\cdot + M1 \blacktriangleleft \quad K_{21} \tag{2.1c}$$

$$M2\cdot + M2 \blacktriangleleft \quad K_{22} \tag{2.1d}$$

The ratios of the rate constants (K_{11}, K_{12}, K_{21}, K_{22}) for a certain radical adding to its own monomer compared to those for the identical radical adding to the other monomer are referred to as the reactivity ratios r_1 and r_2. A value of $r > 1$ indicates that the radical prefers addition to its own monomer, and conversely. Co-polymerizations are categorized based on the values of the product r_1r_2.

When $r_1r_2 = 0$, no monomer will react with its own type, and propagation will solely occur through the addition of the first monomer to the second, resulting in an alternating copolymer.

A more common scenario is $0 < r_1, r_2 < 1$, where both reactivity ratios are below one. When the product of r_1 and r_2 equals 1, copolymerization is deemed perfect, as r_1 equals $1/r_2$, indicating that each radical exhibits an equivalent preference for the other. The arrangement of monomers in the copolymer is arbitrary, determined only by the makeup of the comonomer feed. If one reactivity ratio exceeds unity, the copolymer comprises a greater proportion of the more reactive monomer. As the disparity in reactivity between the two monomers escalates, it becomes increasingly challenging to synthesize copolymers with significant quantities of both monomers.

Procedure

1. A solutions is to be prepared with styrene molar compositions (either of 40, 60, 70, 80 or 100 mol%). Add 20 mL of monomer mixture to the 50 mL round bottom flasks according to Eqs. (2.2) and (2.3).

$$m_{\mathrm{Sty}} = \frac{20 \times x \times \mathrm{Mw(Sty)}}{\frac{(1-x)\times \mathrm{Mw(MMA)}}{\rho_{\mathrm{MMA}}} + \frac{x \times \mathrm{Mw(Sty)}}{\rho_{\mathrm{Sty}}}} \tag{2.2}$$

$$m_{\mathrm{MMA}} = \frac{20 \times (1-x) \times \mathrm{Mw(MMA)}}{\frac{(1-x)\times \mathrm{Mw(MMA)}}{\rho_{\mathrm{MMA}}} + \frac{x \times \mathrm{Mw(Sty)}}{\rho_{\mathrm{Sty}}}} \tag{2.3}$$

where x is the styrene mole% ($0 < x < 1$), m_{Sty} and m_{MMA} are mole fraction of styrene and MMA respectively, Mw stand for the molecular weight and ρ stands for density

2. Further add 20 mg of initiator AIBN to each flask.
3. Since the oxygen present in air may act as an inhibitor, the air above the solution is degassed by bubbling nitrogen for 20 min.
4. Now adjust the temperature of the water baths at 70 °C.
5. After degassing is over, put the flasks in the 70 °C water baths
7. After reacting for twenty minutes, the sample is removed from the water bath and are cooled immediately in an ice bed.
8. The polymer is precipitated by pouring the mixtures into 300 mL methanol.
9. The solid is recovered by filtration on a Buchner funnel.

Result: Copolymer of styrene and methyl methacrylate was prepared by radical polymerization method.

The formation of copolymer of styrene and methylmethacryalate can be confirmed using FTIR and other identification test mentioned in Chap. 1.

Precautions

1. The monomer should be completely washed to remove the inhibitors
2. The time of polymerization should be maintained properly to avoid gelling.

Experiment No. 2.4: Study of the Polymerization Kinetics of Free Radical Chain Polymerization Reaction in Bulk Polymerization System

Aim: To study the polymerization kinetics of free radical chain polymerization reaction in bulk polymerization system.

Apparatus	Chemicals
Thermostatic water bath	Pre washed MMA 60 mL
Digital weighing balance	BPO initiator
Drying oven	Methanol—(nonsolvent 320 mL)
Thermometer (−10 to 110 °C)	4 Nitrogen gas
Beakers, test tubes, measuring cylinder 10 mL	

Theory

Radical chain polymerization is a chain reaction consisting of a sequence of three steps initiation, propagation, and termination. The initiation step is considered to involve two reactions. The first is the production of free radicals by any one of the reactions. The usual case is the homolytic dissociation of an initiator species I to yield a pair of radicals R*.

$$\mathrm{I} \xrightarrow{K_d} 2\mathrm{R}*$$

where k_d is the rate constant for the catalyst dissociation. The second part of the initiation involves the addition of this radical to the first monomer molecule to produce the chain initiating radical M_1*

$$R * + M \xrightarrow{K_d} M_1 *$$

where M represents a monomer molecule and ki is the rate constant for the initiation step. For the polymerization of CH_2=CHY,

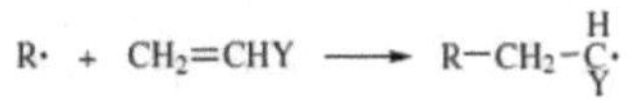

The radical R* is known as an initiator radical or primary radical to distinguish it from the (M_1*) or chain-initiating species. Propagation continues with the growth of M_1* by the successive additions of large numbers (hundreds and perhaps thousands) of monomer molecules. Each addition creates a similar new radical that has the same identity as the previous one, except that it is larger by one monomer unit. The successive additions may be represented by;

$$\mathrm{M}_1 * + \mathrm{M} \xrightarrow{K_d} \mathrm{M}_2 *$$

$$\mathrm{M}_2 * + \mathrm{M} \xrightarrow{K_d} \mathrm{M}_3 *$$

$$\mathrm{M}_3 * + \mathrm{M} \xrightarrow{K_d} \mathrm{M}_4 *$$

etc. etc.

or in general terms

$$\mathrm{M}_n * + \mathrm{M} \xrightarrow{K_d} \mathrm{M}_{n+1} *$$

where k_p is the rate constant for propagation. Propagation with growth of the chain to high polymer proportions takes place very rapidly. After a certain level, the propagating polymer chain stops terminates. Termination of the radicals occur by bimolecular reaction between radicals. Two radicals react with each other by combination (coupling) or, more rarely, by

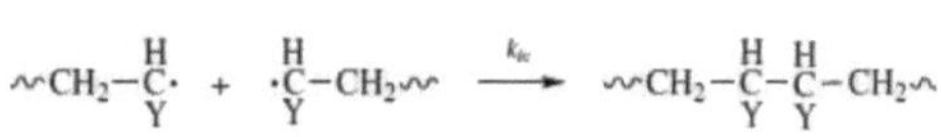

Disproportionation process in which hydrogen radical present at the beta position to one radical center is transferred to another radical center. This result in the formation of two polymer molecules one saturated and one unsaturated

Termination may occur by a combination of coupling and disproportionation. The two different modes of termination can be represented in general terms by

$$M_n{*} + M_m{*} \xrightarrow{K_d} M_{n+m}$$

$$M_n{*} + M_m{*} \xrightarrow{K_d} M_n + M_m$$

where k_{tc} and k_{td} are the rate constants for termination by coupling and disproportionation, respectively. One can also express the termination step by

$$M_n{*} + M_m{*} \xrightarrow{K_d} \text{dead polymer}$$

Rate Expression

Monomer disappears by the initiation reaction as well as by the propagation reactions. The rate of monomer disappearance, which is synonymous with the rate of polymerization, is given by

$$\frac{-\mathrm{d}[\mathrm{M}]}{\mathrm{d}t} = R_\mathrm{i} + R_\mathrm{p}$$

where R_i and R_p are the rates of initiation and rates of propagation, respectively. The number of monomers reacting in the initiation step is far less than the number of monomers in the propagation step in a process of yielding high polymer. For approximation the rate of the initiation step can be neglected and the rate of propagation is only considered in the calculation of the polymerization rate

$$\frac{-\mathrm{d}[\mathrm{M}]}{\mathrm{d}t} = R_\mathrm{p}$$

The rate of propagation approximated to the rate of polymerization, is the sum of many individual propagation steps. Since the rate constants for all the propagation steps are the same, one can express the polymerization rate by

$$R_p = k_p[M*][M]$$

where [M] is the monomer concentration and [M*] is the total concentration of all chain radicals, that is, all radicals of size M_1* and larger. The polymerization rate is not used directly because it comprises of a term for the concentration of radicals. The measurement of the radical concentrations is difficult, as they are very low (~10–8 M), and therefore it is desirable to exclude [M*] from above equation. The steady-state assumption is therefore modified so that the concentration of radicals increases initially, but almost immediately reaches a constant, steady-state value. Very quickly the rate of change of the concentration of radicals becomes and remains zero during the polymerization process. It can also be said that the rates of initiation R_i and termination R_t of radicals are equal or

$$R_\mathrm{i} = R_t = 2k_t[\mathrm{M}^*]^2$$

The steady-state assumption is often used in developing the kinetics of many conventional reactions that involve highly reactive intermediates which are present at very low concentrations conditions. The theoretical validity of the steady-state assumption is validated by experimental observations. The right side of given equation represents the rate of termination. The equation does not specify whether termination is done by coupling or disproportionation reaction as both follow the same kinetic expression. The following reaction can be re arranged as follows:

$$[\mathrm{M}^*] = \left(\frac{R_\mathrm{i}}{2k_t}\right)^{1/2}$$

and substitution

$$R_\mathrm{p} = k_\mathrm{p}[\mathrm{M}]\left(\frac{R_\mathrm{i}}{2k_t}\right)^{1/2}$$

Or

$$R_\mathrm{p} = k_\mathrm{p}[\mathrm{M}]\left(\frac{fk_d[\mathrm{I}]}{k_t}\right)^{1/2}$$

for the rate of polymerization. The above equation describes the dependence of the polymerization rate on the square root of the initiation rate.

Procedure

1. Measure 0.033 g of BPO initiator (use butter paper for weighing) and mix well with 16.5 mL of monomer in a 25 mL beaker. This will be your stock solution of 0.2% (i.e. 0.2 g/100 mL).
2. Take 8 test tubes and pour 5 mL of monomer in each.
3. Mark four test tubes with “T” for time and four test tubes with “I” for initiator concentration.
4. Add 1.0 mL of stock solution in each four test tubes marked with “T”.
5. Add 1, 2, 3, and 4 mL of stock solution in four test tubes marked with “I”.
6. Purge the test tubes with N_2 gas for 3–4 min then plug them with cotton or stopper.

7. Place these test tubes in water bath maintained at 85 °C.
8. Take out 30 mL of methanol in each 8 beakers and cover them with aluminum foil.
9. Withdraw the test tubes marked as "T" from water-bath one by one at elapsed time 35, 40, 45 and 50 min from start of experiment.
10. Slowly pour the content of test tube in methanol (non-solvent) already taken in beaker with mild agitation with glass rod.
11. Filter the precipitated polymer on a pre-weigh and marked (with ball pen) filter paper.
12. Use funnel and conical flask for filtration.
13. Now withdraw all test tubes marked as "I" at 40 min.
14. Repeat the step 10 and 11.
15. Dry these samples in oven maintained at 60 °C for 24 h.
16. Next day weigh the dried polymer along with the filter paper. Subtract the weight of empty filter paper to find the weight of polymer.
17. Report the % yield of polymers w.r.t. polymerization time (T) and initiator concentration [I].
18. Plot the polymerization time (T) versus % yield.
19. Plot the $[I]^{1/2}$ versus % yield.

Observation and Calculations

(A) Data Table: (time vs. % yield)

Volume of monomer (mL) =
Volume of stock solution (mL) =
Total volume of monomer in each test tube (mL) =
Density of monomer (g/cc) =
Weight of monomer (g) =
Water bath's temperature (°C) =

Time (min) t (min)	T (min)				
Weight of polymer (g) P (g)	P (g)				
% Yield = {polymer (g)/monomer (g)} × 100	% Y				

(B) Data Table: ($[I]^{1/2}$ vs. % yield)

Water bath's temperature (°C) =
Polymerization time (min) =

Test tube marking I-1 I-2 I-3 I-4	I-1	I-2	I-3	I-4
Volume of monomer (mL) in test tube				
Volume of stock solution (mL) added				
Total volume (V) of monomer in each test tube (mL)				

(continued)

(continued)

Test tube marking I-1 I-2 I-3 I-4	I-1	I-2	I-3	I-4
Density of monomer = ρ (g/cc)				
Weight of monomer W = V × ρ (g)				
Initiator concentration [I] (g/100 mL) = (W * 100)/V				
I				
Amount of polymer formed P in (g)				
% Yield = (P/W) * 100				

Plot:

(1) Polymerization time (T) versus % Yield.
(2) Sq. root of Initiator Concentration I versus % Yield.

Results and Discussions

Report the result obtained Precautions

1. The monomer should be completely washed to remove the inhibitors
2. The time of polymerization should be maintained properly to avoid gelling
3. The ratio of the two monomers and the reaction conditions should be maintained carefully.

Experiment No. 2.5: Preparation of Nylon 66 Using Interfacial Polymerization Technique

Aim: To Prepare Nylon 66 using interfacial polymerization technique

Apparatus	Chemicals
Beaker	Ethylene diamine
Dropper	Cyclohexane
Glass rod	Adipoyl chloride
	0.5 N NaOH solution

Theory: Interfacial polymerization is a chemical reaction method that produces polymers at the interface of two immiscible liquids or a liquid and air. Step-growth polymerization can be used to create a variety of materials, including: Films, Capsules, Fibers, Separation membranes, and Electrode materials etc. Interfacial polymerization may involve a range of Reaction Schemes, like polycondensation, polyaddition and supramolecular polymerization which are used in sensors, fuel cells, separation/purification membranes, and cargo-loading micro-

and nanocapsules. Interfacial polymerization has extensively been researched and used to prepare polyamides, polyanilines, polyimides, polyurethanes, polyureas, polypyrroles, polyester s, polysulfonamides, polyphenyl esters and polycarbonates.

The most widely used interfacial polymerization methods fall into 3 broad types of interfaces: liquid-liquid interfaces, liquid-in-liquid emulsion interfaces and liquid-solid interfaces. In the liquid-liquid and liquid-in-liquid emulsion interfaces, either one or both liquid phases may contain monomers. Other interface categories, like liquid-gas, solidgas, and solid-solid interfaces may also be used (Fig. 2.3).

Here interfacial condensation polymerization is carried out between a diamine or diol and a diacid, where the two fast-reacting intermediates dissolve in a pair of immiscible liquids, one hydrophobic and the other hydrophillic. Nylon 66 can be prepared by interfacial polymerization technique where two monomers (one in aqueous phase and the other in organic phase) are brought into contact. If hexamethylene diamine and adipoyl chloride are brought in contact directly, the reaction might explode due to its exothermic nature. The monomers are therefore dissolved in aqueous and organic solvents to control the reaction rate.

Procedure

1. Take a 1% solution of Adipoyl Chloride in cyclohexane in a conical flask.
2. Take a 1% solution of Hexamethylene diamine in in 0.5 M NaOH in another conical flask.
3. Now take a small amount of (5 mL) of hexamethylene diamine solution in a beaker with the help of dropper.
4. Add the organic solution of Adipoyl chloride in cyclohexane on the hexamethylene solution using a dropper.
5. Within a few seconds the mixture will turn cloudy. No with a help of a glass rod, pull the polymer (nylon 66) formed at the interface slowly.
6. More the polymer at interface is taken out, more and more monomers will come in contact and more polymer will be formed till one of the monomer is exhausted.
7. Collect the polymer thread.

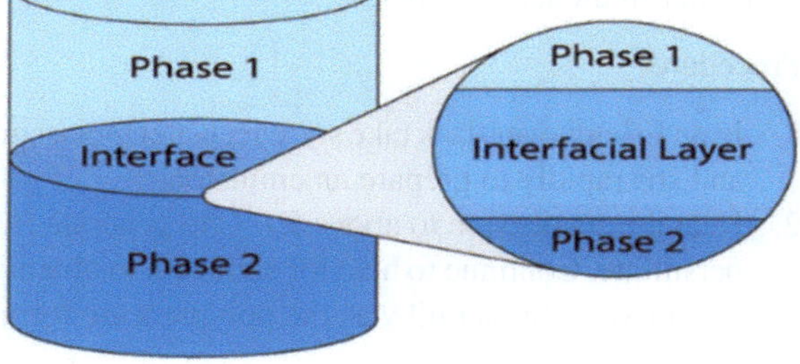

Fig. 2.3 An experimental set up for interfacial polymerization

Observations

Amount of Nylon 66 obtained is .... g

Amount of Adipoyl Chloride Taken=

Amount of Hexamethylene diamine =

Percentage yield=

Result: Nylon 66 is prepared with ......% yield.

The formation of nylon 66 can be confirmed using FTIR and other identification test mentioned in Chap. 1.

Precautions

1. The thread of the Nylon 66 should be drawn very carefully. It should be taken out so that new monomers can come in contact.

Experiment No. 2.6: Prepare Polystyrene Latex Using Emulsion Polymerization

Aim: To prepare Polystyrene latex using emulsion polymerization

Apparatus	Chemicals
Magnetic stirrer/hot plate	7.1 g styrene
Erlenmeyer flask 100 mL with a magnetic stir bar	12.8 g Water
	3.1 mL of a 0.68% (0.34 g in 50 mL distilled water) solution of potassium persulfate
	0.356 g SDS in 10 mL water

Theory: Emulsion polymerization of styrene is a process in which the dispersion of water-insoluble styrene monomers is used as the monomer particles suspended in an aqueous phase. The polymerization process typically involves three stages: (a) nucleation of particle nuclei, (b) growth of latex particles, and c) depletion of residual monomer in the latex particles.

Procedure

1. In an Erlenmeyer flask take styrene, water, potassium persulfate and soap solution and stir rapidly to prepare an emulsion.
2. Heat the suspension to above 50 °C to generate free radicals from the potassium persulfate. Continue to heat for about 2 h to obtain a reasonable yield of polymer.
3. The mixture turns milky as the polymers are formed.

Observation: Milky solution of polystyrene latex is obtained.

Result: Polystyrene latex is prepared by emulsion polymerization technique.

The formation of polystyrene can be confirmed using FTIR and other identification test mentioned in Chap. 1.

Precautions:

1. Maintain appropriate temperature and mixing conditions,
2. Prevent contamination and coagulation.

Experiment No. 2.7: Preparation of Polyacrylonitrile by Solution Polymerization Technique

Aim: To prepare polyacrylonitrile by solution polymerization technique

Apparatus	Chemicals
Round bottom flask (500 mL)	Ammonium peroxide sulphate (APS)
Autoclave	Acrylonitrile (AN)
Nitrogen cylinder, filter paper	Hydrochloric acid 37%, dimethyl formamide, and distilled water

Theory: The solution polymerization procedure is used in industrial polymerization. In this case, a monomer and a catalyst or initiator are mixed together and dissolved in a non-reactive solvent.

The final polymer is also soluble in the solvent of choice. The reaction rate is decreased when the solvent absorbs the heat released by the process. Additionally, autoacceleration at high monomer concentrations is prevented by decreasing the viscosity of the reaction mixture. Dilution also aids in heat transmission, which is one of the primary issues with polymer synthesis because most polymerizations are exothermic events. After the necessary conversion is accomplished, excess solvent must be removed to obtain the pure polymer.

Procedure I

1. Acrylonitrile was polymerized in a 500 mL autoclavable container and agitated in a nitrogen atmosphere using APS as an initiator and HCl as an activator of the polymerization process in aqueous media, at 80 °C for 6 h.
2. In a batch containing different monomer ratios (82, 90, 98, 110, and 134) g/L water, at different pH (2.34, 2.24, and 2.18) and different concentrations of APS (0.1, 0.09, 0.07, 0.05, and 0.03) g/L, at 80 °C for 6 h) to obtain different molecular weights of PAN.
3. The polymers were filtered, washed successively with distilled water and acetone, and dried under vacuum at 60 °C till a constant weight was obtained.

Procedure II

1. Into 250 cc round-bottomed flask 175 cc water is added and nitrogen is babbled for about 30 min.
2. Then add 15 cc acrylonitrile, 0.5 cc of $Na_2S_2O_5$ solution (5% in water) and 2.5 cc of $FeSO_4{\cdot}7H_2O$, solution (10 mg in 100 cc water + 2 cc conc. H_2SO_4) are introduced and subsequently 2.5 cc of $K_2S_2O_8$ solution (5% in water) is added and mixed briefly at 20 °C.
3. The mixture is stirred at 20 °C for about 1 h, the precipitated polymer is filtered washed with water then with methanol and dried at 50 °C in vacuum

Calculation

Monomer Conversion: The evidence of the copolymerization process was investigated by varied duplicate experiments. The conversion yield of prepared PAN was calculated according the following equation 1.

$$\text{Monomer conversion} = \frac{\text{Weight of polymer(g)}}{\text{Monomer feed(g)}} \times 100$$

Observation: PAN Polymer solution is obtained.

Results: PAN is prepared by solution polymerization technique

The formation of polyacrylonitrile can be confirmed using FTIR and other identification test mentioned in Chap. 1.

Precautions

1. Maintain inert conditions, manage heat release
2. Handle flammable materials safely.

Experiment No. 2.8: Preparation of a Copolymer of VINYL ACETATE and Butyl Acrylate by Emulsion Polymerization Method

Aim: To prepare a copolymer of vinyl Acetate and Butyl Acrylate by emulsion polymerization method

Apparatus	Chemicals
Magnetic stirrer and heating mantle	Poly(vinyl alcohol)
Glass plate	Vinyl acetate
Nitrogen cylinder	Butyl acrylate
	Sodium lauryl sulphate, ammonium persulphate

Theory

Emulsion polymerization of vinyl acetate and butyl acrylate involves creating a stable dispersion of polymer particles in water, using these two monomers. The process typically involves a free radical polymerization mechanism where the monomers are emulsified in water with a surfactant, and then initiated by a water-soluble initiator. The resulting latex, a milky-white dispersion, is used in various applications like adhesives, coatings, and construction materials.

Procedure

1. Sodium lauryl sulphate (4 g) and Poly(vinyl alcohol) (4 g) are added to 100 mL of water in the reaction vessel.
2. The mixture is stirred and heated to 80 °C to dissolve the emulsifiers and nitrogen is purged through the system to remove and oxygen.
3. Prepare a monomer mixture of 100 g of vinyl acetate and 0/5/10 g of butyl acrylate.
4. Add 0.8 g of ammonium persulphate to the reactor and immediately begin addition of the monomer mixture at about 5 cm^3/min.
5. Maintain the reaction temperature at 80–85 °C during the addition and for 30 min. after completion of addition.
6. After completion of reaction, transfer to a glass container and allow the emulsion to cool to room temperature.
7. The solid content of the emulsion and film properties should be tested by casting and accurately weighing 1.0 g on a clean glass plate and drawing at 80 °C. The dry film is weighed and solids content determined. The film should be peeled from the plate and examined.

Observation: Polymer film is obtained.

Results: The copolymer of vinyl acetate and butyl acrylate is prepared. The yield of the polymer is ……%.

The formation of copolymer of vinyl acetate and butyl acrylate can be confirmed using FTIR and other identification test mentioned in Chap. 1.

Precautions

1. Maintain appropriate temperature and mixing conditions,
2. Prevent contamination and coagulation.

Experiment No. 2.9: Atom Transfer Radical Polymerisation (ATRP) of Methyl Methacrylate

Aim: To Prepare Methyl Methacrylate using Atom Transfer Radical Polymerisation (ATRP)

Apparatus	Chemicals
Round bottom flask	Methyl methacryate (MMA) purified
Magnetic stirrer	Methyl-2-bromopropionate
Nitrogen cylinder	CuBr
Rota evaporator	N,N,N′,N′,N″-Pentamethyldiethylenetriamine (PMDETA)
	Methanol, dichloromethane

Theory

Atom Transfer Radical Polymerization (ATRP) is a controlled radical polymerization technique used to synthesize polymers like PMMA (Polymethyl Methacrylate) with well-defined molecular weights and architectures. In ATRP, a metal catalyst (e.g., copper based) reversibly activates a dormant polymer chain, allowing it to grow by adding monomer units. This controlled process enables the creation of polymers with specific properties and structures, including block copolymers and star polymers.

Mechanism: ATRP involves a dynamic equilibrium between dormant and active polymer chains, controlled by the catalyst system. This equilibrium allows for controlled chain growth and prevents termination reactions, leading to polymers with narrow molecular weight distributions.

PMMA synthesized via ATRP can be used in various applications, including the creation of block copolymers, star polymers, and materials with specific properties (Scheme 2.3).

Procedure

1. To perform ATRP of MMA, CuBr (175 mg, 1.2×10^{-3} mol) and methyl-2-bromopropionate (MBP) (0.130 ml, 1.22×10^{-3} mol) were added to a 100 mL Schlenk round bottom flask equipped with a magnetic stirrer bar.
2. The flask was degassed and back filled with nitrogen three times before introducing deoxygenated MMA (12.8 mL, 0.120 mol) and PMDETA (0.25 mL, 1.22 $\times 10^{-3}$ mol) via purged syringe.
3. After the addition of PMDETA the solution becomes green due to catalyst-ligand complex formation and the flask was placed in an oil bath at 90 °C for 40 min.
4. The above reaction product was dissolved in 50 mL of DCM and precipitated in tenfold excess of methanol with constant stirring and filtered.
5. It is redissolved in DCM and reprecipitated in excess amount of ether to remove the catalyst and concentrated using rotavapor then dried in vacuum.

GENERAL MECHANISM

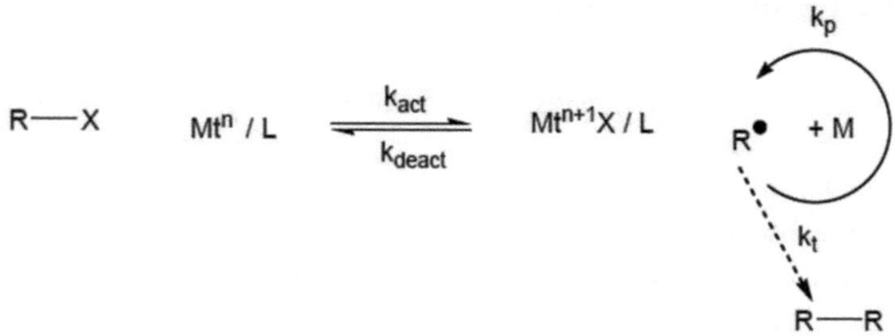

Figure 1 – Mechanism of ATRP

REACTION INVOLVED

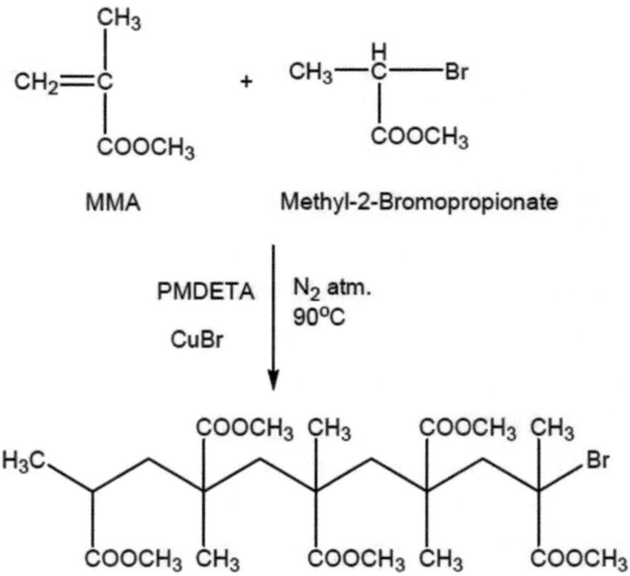

Scheme 2.3 ATRP polymerization reaction of PMMA

Result: White Color precipitate is obtained. The yield of the polymer is …..%

The formation of copolymer of vinyl acetate and butyl acrylate can be confirmed using FTIR and other identification test mentioned in Chap. 1.

Precautions

1. Meticulous purification of all reaction components
2. Proper deoxygenation of the reaction mixture
3. Appropriate catalyst selection, and careful solvent choice.

Experiment No. 2.10: Prepare Styrene Butadiene Rubber by Emulsion Polymerization Technique

Aim: To prepare Styrene Butadiene Rubber by emulsion polymerization technique

Apparatus	Chemicals
450 mL Parr reactor	The purified monomers: styrene, butadiene
High-pressure pump	Initiator 2,2′-azobis(2-methyl butyronitrile) and tert-dodecylmercaptan
Homogenizer	Hexadecane, octadecyl acrylate
Reactor	Oleic acid, potassium hydroxide

Theory: Same as Expt No. 2.6.

Procedure

1. A pre-formed styrene miniemulsion (as described in Table 2.1) is added to the Parr reactor at 5 °C under stirring condition (100 rpm) for fifteen minutes to reach thermal equilibrium.
2. Now liquid butadiene (65 g or 102 mL) (density of butadiene is given in Table 2.2) was added via a high-pressure pump and homogenized for thirty minutes at 200 rpm.
3. Then, the reactor was heated to 70 °C ± 2, at 350 rpm and the polymerization was done overnight at around 8 bar pressure.

Density of Butadiene at 5 °C = 0.64 g/cc

Note: Additional studies may be done by varying the amount of different chemicals used in the formulation.

Result: Milky white emulsion of SBR is obtained

The formation of SBR can be confirmed using FTIR and other identification test mentioned in Chap. 1.

Precautions

1. Maintain appropriate temperature and mixing conditions,

Table 2.1 Formulation of preformed styrene mini-emulsion

Chemicals	Mass (g)	wt% based on styrene
Styrene	30	–
Hexadecane or octadecylacrylate	1.2	4
Deionized water	100	–
Potassium hydroxide	0.3	–
Oleic acid	0.9	3
Tert-dodecylmercaptan	1.2	4

Table 2.2 Formulation of the SBR miniemulsion

Chemicals	Weight (g)	wt% based on organic phase
Butadiene	65	–
Styrene	30	–
Hexadecane or octadecylacrylate	1.2	1.2
Deionized water	100	–
Tert-dodecylmercaptane	3.8	4
Potassium hydroxide	0.3	
Oleic acid	0.9	1
VAZO 67	1.9	2

2. Prevent contamination and coagulation.

Experiment No. 2.11: Preparation of Polythiokol/ Polysulphides

Aim: To Prepare Polythiokol/polysulphides.

Apparatus	Chemicals
6 inches long (15 cm) 2 10 mL vials with Teflon cap liners	Sodium hydroxide solution
Hot plate Chemical resistant gloves	1 M sulfur, 1,2-dichloroethane
10 mL graduated cylinder glass pipette (dropper)	Distilled or deionizer water
2400 mL beakers	Copper wire

Theory

The characteristics of the polymers are contingent upon the length of the aliphatic chains and the quantity of sulfur atoms present. The polymer exhibits elastomeric properties when four or more sulfur atoms are present per monomer, but lesser quantities of sulfur result in a rigid, brittle resin.

Key features include:

(a) Thiokol exhibits resistance to the effects of oxygen and ozone.
(b) It is also impervious to the effects of petroleum lubricants and organic solvents.
(c) Thiokol exhibits exceptional resistance to swelling from organic solvents; nonetheless, benzene and its derivatives induce some swelling.
(d) Thiokol films exhibit significant impermeability to gases.

Polythiokol has significant applications such as:

- It is utilized in hoses and tank linings for the management and storage of oils and solvents.
- It is utilized in the lining of vessels employed in chemical manufacturing.
- It is applicable to engine gaskets and similar items that interact with oil.
- Thiokol combined with oxidizing substances serves as a propellant in rocket engines.

The synthesis of Thiokol rubber involves a two-step procedure. The initial stage involves the synthesis of sodium polysulfide through the interaction of sulfur (S_8) with a strong basic, sodium hydroxide (NaOH).

$$S8 + 2NaOH \rightarrow Na_2S_8$$

At room conditions, sulfur is normally in the form of S8 rings and chains. The reaction with sodium hydroxide produces a mixture of mostly Na_2S_8 with other chain lengths of sulfur present. That mixture is called sodium polysulfide.

The second step of this preparation is the reaction of sodium polysulfide with ethylene dichloride.

$$Cl{-}CH_2{-}CH_2{-}Cl + Na_2S_x \rightarrow [-CH_2{-}CH_2{-}S_8{-}]_n + 2nNaCl$$

Procedure

1. In a beaker containing 100 mL. of dist. Water dissolve 5 g. of sodium hydroxide and heat to boiling.
2. Add 7–10 g. of sulfur flowers, stirring until all of the sulfur has dissolved. A deep red solution is obtained due to the formation of sodium polysulphide. If some sulfur remains undisclosed the solution is filtered.
3. Allow the solution to cool to below 83 °C, the boiling point of 1,2-dichloroethane, and add 20 mL. of 1,2-dichloroethane with stirring.
4. Continue to stir for a few minutes (15–20 min.) while a rubber polymer separates out as a lump.
5. Decant the supernatant liquid and wash the product several times with water and leave in the fume cupboard for a few minutes to allow excess 1, 2-dichloroethane to evaporate.

Results: Weight Polythiokol resin = W g

The formation of polythiokol can be confirmed using FTIR and other identification test mentioned in Chap. 1.

Precautions

1. Handle sodium hydroxide and 1,2-dichloroethane carefully.

Experiment No. 2.12: Preparation of Hydrogel

Aim: To synthesize hydrogels based on 2-hydroxyethyl(methacrylate) and to study the effect of cross-linker ratio on degree of swelling.

Apparatus	Chemicals
Syringe	2-Hydroxyethyl(methacrylate), de-ionized water
Nitrogen cylinder	PMDETA (N,N,N′,N″,N″-pentamethyldiethylenetriamine)
	N,N′-Methylene bis(acrylamide)
	Ammonium peroxysulfate (APS)

Theory

Hydrogels are three dimensional network structure of hydrophilic monomers. Due to the hydrophilicity of the monomers, the three dimensional network structure can exhibit swelling behavior. Hydrogels can absorb a large quantity of water in their network without disturbing their 3D structure. This results in flexibility and swelling properties of the hydrogel structures. Common applications of hydrogels are in tissue engineering, cell culture, drug delivery, agriculture, as biosensors, for contact lenses, for encapsulation of quantum dots, electrocardiography (ECG) medical electrodes etc. (Scheme 2.4).

Procedure

- 2-Hydroxyethyl(methacrylate) monomer and methylenebis(acrylamide) are taken in a 3 mL syringe and mixed followed by the addition of DI water (1.5 g) for dissolution of the contents.
- Then the mixture is purged with N_2 gas.
- Now add PMDETA to this solution.
- Add the redox initiator APS and the solution is mixed properly.
- Now the syringe is sealed using paraffin.
- The solution is now left at room temperature for 20 min.
- The gel is now removed from the syringe, washed with water (100 mL) for 10 h which included periodic changes of the solvent every 2 h.

For studying the effect of crosslinker, following compositions may be used (Table 2.3).

Determination of swelling degree

- Use the gravimetric method to estimate the swelling kinetics of the hydrogels. Weigh the dried hydrogel (M_0).
- After being immersed in DI water at room temperature for predetermined time intervals, clean the sample with filter paper to remove excess water on the surface and weigh the hydrogel (M_t).
- All samples should be measured minimum three times.

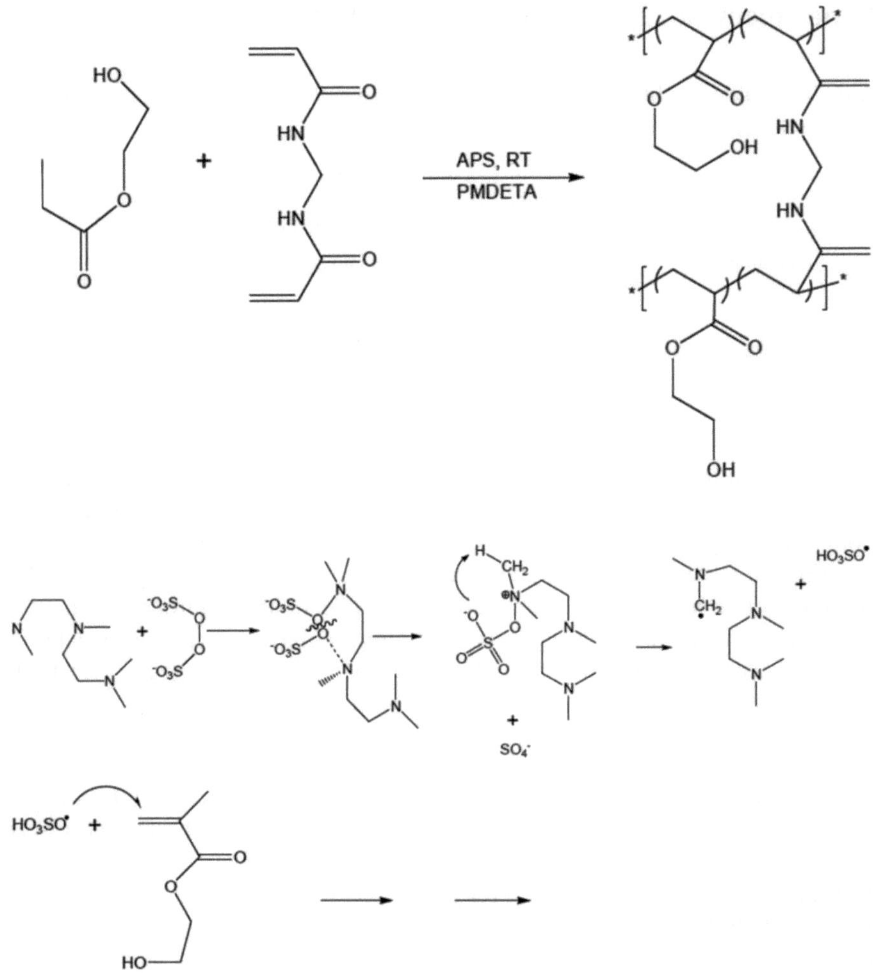

Scheme 2.4 Reaction of hydrogel Preparation [based on 2-hydroxyethyl(methacrylate)]

Table 2.3 The ratio of the reagents for preparation of hydrogels

Sl. No.	2-Hydroxyethyl methacrylate (mmol)	Methylene bisacrylamide (mmol)	Ammonium peroxysulfate (μmol)	PMDETA (μmol)
1	1.7	0.085	34	64
2	1.7	0.17	34	64
3	1.7	0.255	34	64
4	1.7	0.340	34	64

The swelling degree is calculated according to the following equation:
Swelling % = (Ws - Wd) / Wd × 100

Observation Table

Sample No.	Hardness	Swelling degree
1		
2		
3		
4		

Result

A hydrogel is obtained with degree of swelling …..

Precautions

1. Maintain sterile conditions,
2. Control reaction parameters like temperature and pH,
3. Handle crosslinking agents carefully.

Chapter 3
Preparation of Thermosets

Abstract Thermosets are the polymeric structure which are three dimensional in nature. These structures are rigid in nature and cannot be reshaped by any means like application of heat. Among various thermosets, phenol formaldehyde resin, urea formaldehyde resin, and melamine formaldehyde resin are important. These resins are prepared by condensation method in multiple stages. In this chapter the preparation methods for all these resins are discussed and described.

Keywords Polymerization · Urea formaldehyde · Melamine formaldehyde · Phenol formaldehyde · Epoxy

Thermosets are a class of polymeric materials that form a three-dimensional network structure. Thermosets, which are typically crosslinked, give polymer chains their stiffness. The chains cannot move or slide over one another because they are so intertwined.

Unlike thermoplastics, these polymers cannot be melted and remolded by heating. Compared to thermoplastic materials, thermosetting plastics are more stiff because of their three-dimensional network. Due to their ability to hold their shape without experiencing chemical breakdown, they are used in high-temperature applications. More aromatic content and a higher crosslink density make a thermoset more resilient to heat breakdown and chemical attack. As crosslink density increases, so do mechanical strength and hardness. They usually disintegrate before melting.

Hard thermosets exhibit permanent plastic deformation under load, in contrast to rubbers. Generally, once cured, thermoset polymers or elastomers cannot be melted and reshaped. Thermosets cannot be recycled because of this.

Condensing two distinct liquid or semirigid monomers or prepolymers results in a condensation polymer, which is a common thermoset. The growth of thermosets typically occurs in three stages. Dimers or trimers are formed in the first stage. These dimers/trimers subsequently condense to create a viscous resin in the second stage.

When additional curing agent or crosslinker is added, this resin creates a stable, stiff, three-dimensional network structure.

S. D. Sadhu et al., *Source Book of Experimental Techniques and Analysis of Polymeric Materials*, https://doi.org/10.1007/978-981-92-2088-5_3

The widely accessible thermosets are covered here. These three types of formaldehyde resin are urea formaldehyde, phenol, and melamine.

Phenol formaldehyde resin is utilized in adhesives, coating applications, and the creation of molded goods. The most common uses for phenol formaldehyde are as adhesives for plywood and wood. It is also applied on cement, wood, and other materials. It can also be used to laminate decorative items. As a nonconducting substance, this resin is used in electrical appliances.

There are numerous uses for urea formaldehyde resin, including coatings and adhesives. These can once more be applied as pressed wood adhesive, plywood, and wood. UF resins can be used to coat metal, wood, and fabrics. UF resin is used in wrinkle-resistant textiles, cotton blends, foam artificial snow, etc. Urea formaldehyde resin is used in agriculture to regulate the release of nitrogen to promote plant growth. UF resin usually has high flexural modulus, high thermal distortion temperature, and high tensile strength. Additionally, melamine formaldehyde resin finds employment in textile, automotive, and construction industries. These are utilized in wall coverings, particle boards, and laminate flooring. Surface coating for improved surface finish is a common use. Additionally, MF resin is utilized in textile finishes, fireproof fleece, and protective apparel. Composite applications, dishware, tableware, adhesive and bonding applications, etc., are further significant uses.

Experiment No. 3.1: Prepare Phenol-Formaldehyde Resin (Novolac)

Aim: To prepare phenol-formaldehyde resin (Novolac)

Apparatus	Chemicals
Round bottom flask	Phenol (2 g)
Beaker	40% aq formaldehyde solution or formalin (2.5 cc)
Glass rod	Glacial acetic acid (5 cc)
	Conc. HCl (8 cc)

Theory: Phenol formaldehyde resin is used in coating applications. Phenol formaldehyde resin, also known as P-F resin or phenolic resins (referred to as phenoplasts), is a significant category of polymers synthesized through the condensation polymerization of phenol and formaldehyde either in acidic or in alkaline environments.

Initially, methylol phenol derivative is formed depending upon the phenol to formaldehyde ratio. In the next step, the phenol formaldehyde derivatives react with one another or with phenol to produce a linear polymer (Novolac) or a more extensively crosslinked polymer (Bakelite). Bakelite, a highly crosslinked thermosetting polymer can be produced through the further condensation of novolac or its methylol derivative. The crosslinking may be done by adding hexamethylene tetramine during synthesis (Scheme 3.1).

1. 5 cc of glacial acetic acid and 2.5 cc of 40% aq formaldehyde solution are taken in a 100 cc beaker along with 2 g phenol. To this mixture, conc. HCl is added dropwise with vigorous stirring (preferably in a magnetic stirrer), and wrap it by a glass rod till a gummy mass appears.
2. Wash the residue several times with to make it free from acid.
3. Filter the product and weigh it after drying in folds of a filter or in an oven. Report the yield of polymer formed.

Observation

Weight of polymer formed = [Weight of the polymer formed + the weight of the container] − Weight of the container in g.

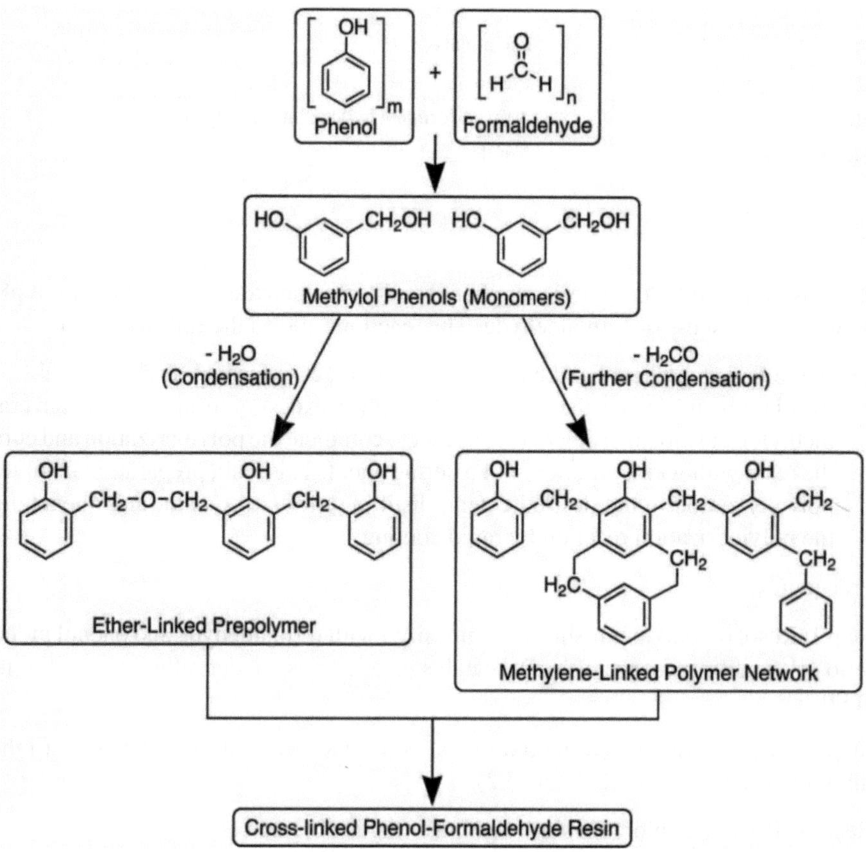

Scheme 3.1 Preparation of phenol formaldehyde resin procedure

Results: Weight of phenol formaldehyde resin = W g

The formation of phenol formaldehyde can be confirmed using FTIR and other identification tests mentioned in Chap. 1.

Precautions

1. Handle phenol and formaldehyde; they can cause irritation and are potential carcinogens,
2. ensure proper ventilation and personal protective equipment to prevent inhalation of harmful vapor.

Experiment No. 3.2: Prepare Resol Using Polycondensation Reaction

Aim: To prepare Resol using polycondensation reaction

Apparatus	Chemicals
Round bottom flask	Phenol
Beaker	40% aq formaldehyde solution or formalin
Heating plate	NaOH

Theory

Resol is prepared by polymerizing phenol and formaldehyde at basic (alkaline) pH with a molar excess of formaldehyde. The reaction follows the following steps:

1. In the first step phenol and methylene glycol react to form methylol phenol.
2. Resol is obtained when the resin is prepared in an excess of formaldehyde. Several methylol and dibenzyl ether groups react to complete the polymerization and cure the resin without incorporation of a curing agent. This resin preparation involves polymerization to the desired extent, distilling off excess water and quenching the polymerization reaction by rapid cooling.

Procedure

Step 1: Resol prepolymer resin was synthesized with formaldehyde and phenol molar ratio 1:1.8 in the presence of NaOH catalyst in a RB flask under refluxing conditions at 90 °C.

Step 2: The resulting reaction mixture was stirred and heated to 90–95 °C until the mixture became cloudy (Scheme 3.2).

Results: Rose-red colored resin weighing ….. g is obtained.

The formation of Resol can be confirmed using FTIR and other identification tests mentioned in Chap. 1.

Scheme 3.2 Preparation of resole observation: rose-red colored resin is obtained

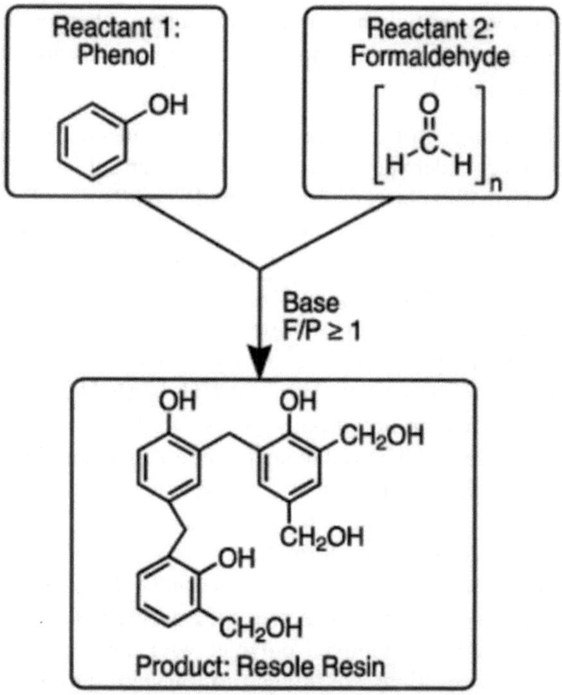

Precautions

1. Handle phenol and formaldehyde; they can cause irritation and are potential carcinogens,
2. ensure proper ventilation and personal protective equipment to prevent inhalation of harmful vapors.

Experiment No. 3.3: Prepare Urea Formaldehyde Resin

Aim: To prepare urea formaldehyde resin

Apparatus	Chemicals
Beaker	Phenol (2 g)
Round bottom flask	40% aq formaldehyde solution or formalin (2.5 cc)
Heating plate	Ammonium hydroxide solution

Theory: Urea formaldehyde (UF), or urea-methanal, is an opaque thermosetting resin synthesized from urea and formaldehyde. These resins are utilized as adhesives, plywood, particle board, medium-density fiberboard (MDF), and molded products. In agriculture, urea formaldehyde compounds are among the most prevalent slow-release fertilizers.

Reaction chemistry: The primary constituents for amino resins are urea, melamine, and formaldehyde. Formaldehyde is often sold in a 37% aqueous solution referred to as formalin. Gaseous formaldehyde rapidly dissolves in water, reacting to generate methylene glycol: $HCHO + H_2O \rightarrow HO\text{–}CH_2\text{–}OH$. The other principal variant of formaldehyde is solid paraformaldehyde (flake or powder), including a mixture of poly(oxymethylene glycols), $HO(CH_2O)_nH$, where n = 8–10. These chains easily regenerate formaldehyde when mixing with water. UF and MF condensations include two primary processes. Hydroxymethylation can be catalyzed by either an acid or a base. Hydroxymethylation: $R\text{-}NH_2 + HCHO \rightarrow R\text{-}NH\text{–}CH_2\text{–}OH$. This typically constitutes the initial phase in the synthesis of UF polymers, with these methylol molecules synthesized in neutral or alkaline conditions. The second significant reaction is the production of a methylene bridge, which is exclusively catalyzed by acid.

Methylene bridge formation: $R\text{-}NH\text{–}CH_2OH + H_2NR{\cdot}RNH\text{–}CH_2\text{–}NHR + H_2O$. This reaction leads to the production of dimers, polymer chains, or networks. The stable resin intermediate produced during hydroxymethylation can be transformed into a solid crosslinked polymer with the incorporation of an acid catalyst. Two primary reactions are thought to transpire during crosslinking. The initial process is the creation of dimethylene ether linkages, represented as $2RNHCH_2OH \rightarrow RNHCH_2\text{–}O\text{–}CH_2NHR + H_2O$, while the subsequent reaction is a rearrangement that results in the development of a diaminomethylene bridge, denoted as $RNHCH_2\text{–}O\text{–}CH_2HNR \rightarrow RNHCH_2NHR + HCHO$. The resultant matrix is seen below for melamine-formaldehyde. The four fundamental reactions significantly influence the reaction chemistry of UF and MF resins, and their meticulous regulation remains a focal point of rigorous industrial and academic investigation (Scheme 3.3).

Procedure I (Base catalyzed)

Urea formaldehyde resin is prepared by reacting urea and formaldehyde in a three-step process:

1. Urea and formaldehyde are reacted in the presence of ammonia, with a formaldehyde to urea ratio of 1.2–1.8. The pH must be maintained above 7.
2. The mixture is then heated to 85–95 °C for about two hours to form a cyclic triazone/triazine polymer.
3. The cyclic polymer is then heated further to form a thermosetting polymer.

Procedure II (Acid catalyzed)

1. Pour about 5 mL of 40% formaldehyde solution in a 100 mL beaker and add about 2.5 g of urea with constant stirring till the solution is saturated. Now a few drops of conc. H_2SO_4 are added with constant stirring.
2. As white solid mass appears in the beaker, wash the white solid with water and dry it in the folds of filter paper.

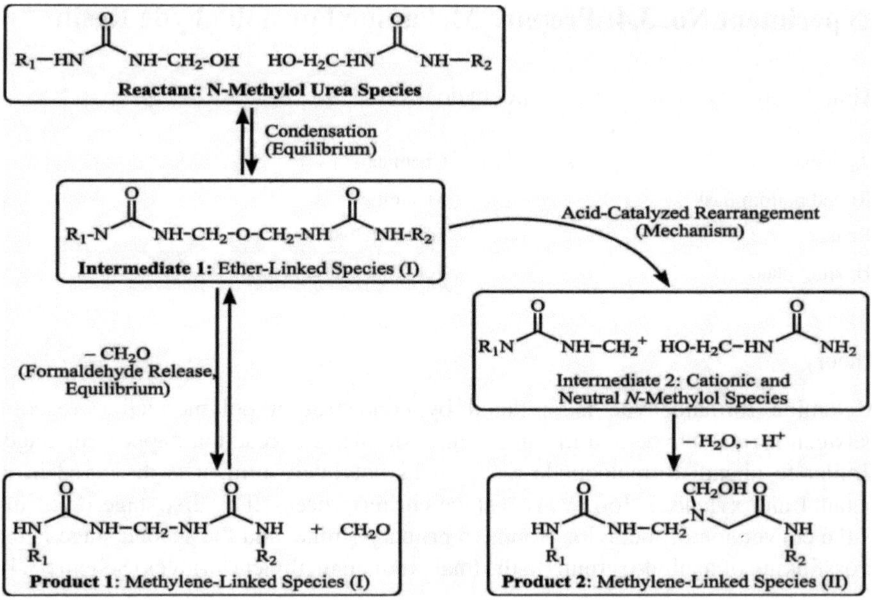

Scheme 3.3 Preparation of urea formaldehyde

3. Weight the yield of product and report.

Observation

Mass of the beaker (W1) = ---------g.

Mass of the beaker with urea formaldehyde (W2) = --------g.

Therefore mass of urea formaldehyde (W2 − W1) = --------g.

Result

The yield of urea formaldehyde = -------g.

The formation of urea formaldehyde resin can be confirmed using FTIR and other identification tests mentioned in Chap. 1.

Precautions

1. While adding concentrated H_2SO_4, it is better to stay a little away from the beaker since the reaction sometimes becomes vigorous.
2. The reaction mixture should be stirred continuously.
3. Proper ventilation and personal protective equipment are essential due to formaldehyde's volatility and potential to cause respiratory irritation, skin rashes, and even allergic reactions.

Experiment No. 3.4: Prepare Melamine Formaldehyde Resin

Aim: To prepare melamine formaldehyde resin

Apparatus	Chemicals
Round bottom flask	Melamine
Beaker	Formaldehyde
Heating plate	Tetrahydrofuran, NaOH

Theory

Melamine formaldehyde is prepared by condensation polymerization reaction between formaldehyde and melamine in basic pH. The reaction happens in stages similar to phenol formaldehyde resin. In its butylated form, it is dissolved in n-butanol and xylene. It follows two-stage curing process. The first stage of curing is the conversion of methylol groups to primary amine, and the second stage is the crosslinking of methylol groups to the final product, methylene bridges (Scheme 3.4).

Procedure

1. A low methanol-37% formalin solution (Note 1, 2500 g, 30.9 mol) is introduced into a 5 L three-necked flask fitted with a stirrer, reflux condenser, and thermometer. The pH of the formalin is modified to 8.0–8.2 using 2 N NaOH, requiring approximately 3.8 mL.
2. Melamine (1890 g, 15 mol) is introduced with rapid stirring. The slurry is heated to reflux over 30–40 min, ensuring a consistent temperature increase.
3. The mixture is refluxed at temperatures ranging from 98 to 102 °C for a duration of 10 min.
4. The heating mantle is detached, and the resin syrup is uniformly cooled to 80–85 °C within 10–15 min using a cold-water bath.
5. The resin is held at this temperature for 1–2 h until the reaction concludes.

Observation

Mass of the beaker (W1) = ---------g.

Mass of the beaker with melamine formaldehyde (W2) = --------g.

Therefore mass of melamine formaldehyde (W2 − W1) = --------g.

Result

The yield of melamine formaldehyde = -------g.

The formation of melamine formaldehyde can be confirmed using FTIR and other identification tests mentioned in Chap. 1.

Precautions

1. Manage formaldehyde exposure, controlling reaction conditions.

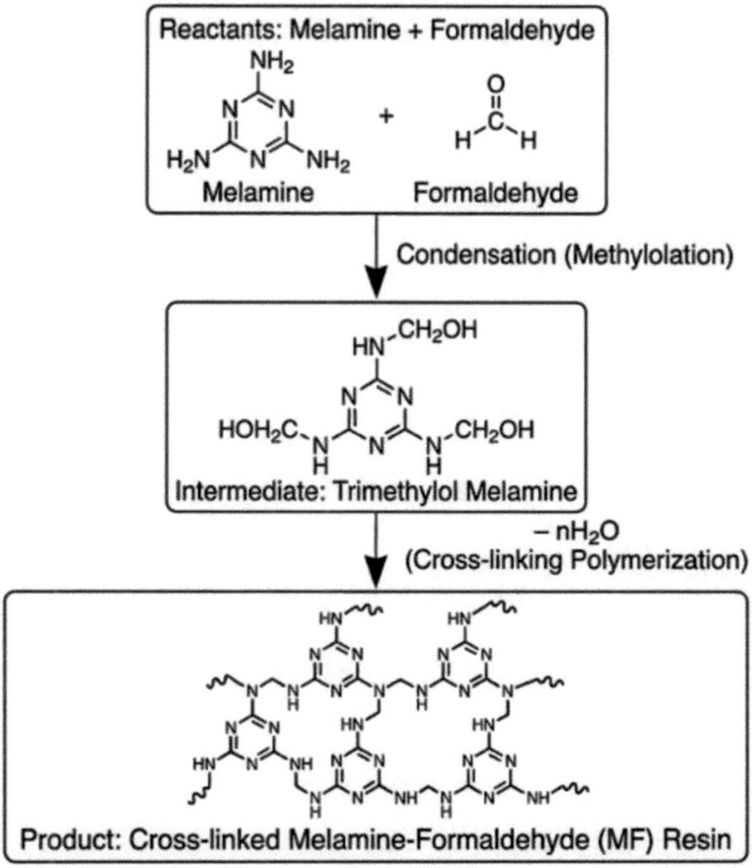

Scheme 3.4 Preparation of melamine formaldehyde resin

Chapter 4
Synthesis of Conducting Polymers

Abstract Conducting polymer is a class of polymer which can carry electricity. These are a special group of polymers which contain alternate conjugated double bonds. The most simplest conducting polymer structure is polyacetylene. Among other conducting polymers polyaniline, polythiophene and polypyrrole are important ones. These conducting polymers require to be dopes with ions so that the conduction of charge may happen. In this chapter the preparation and properties of some of the commercially important conducting polymers are reported.

Keywords Conducting polymer · Polyaniline · Polypyrrole · Polythiophene

Conducting polymers are a special class of polymers which can conduct electricity. Conventional polymers are well known as a class of materials which are completely insulating in nature. The invention of conducting polymers, particularly polyacetylene, revolutionized materials science, paving the way for a wide range of applications. This discovery, made by Alan J. Heeger, Alan G. MacDiarmid, and Hideki Shirakawa in the 1970s, led to their Nobel Prize in Chemistry in 2000. The key to making these polymers conductive is doping, a process of introducing impurities (dopants) to alter the polymer's electronic structure. This process creates a charge carrier, enabling the flow of electricity. The discovery of polyacetylene is then followed by a series of developments and inventions like.

Conducting polymers have π-conjugated polymer backbone with alternating single and double (or triple) bonds. Theses conjugated systems can transport charges, independent of their intrinsic conductivities (conductor or semiconductor) and charge transport characteristics. The chemical structure of polyacetylene is energetically more stable when the polymer chain has a bond length alternation =0. This results into an energy gap (E_g) between the highest valence band and lowest conduction band (Fig. 4.1), resulting in insulating behavior.

Polyacetylene is the first conducting polymer invented. It is a polymer of acetylene (C_2H_2). On polymerization, polyacetylene forms which has alternate double bonds. In case of a charge development at any point in the structure the charge gets distributed over the whole polymer chain through resonance. Overlapping of such

S. D. Sadhu et al., *Source Book of Experimental Techniques and Analysis of Polymeric Materials*, https://doi.org/10.1007/978-981-92-2088-5_4

conjugated π electrons over the entire backbone results in the formation of valence band and conduction band separated by a significant band gap. Activation of polymer by thermal or photolytic effect gives electrons sufficient energy to jump the gap and reach into lower levels of the conduction band making it conducting.

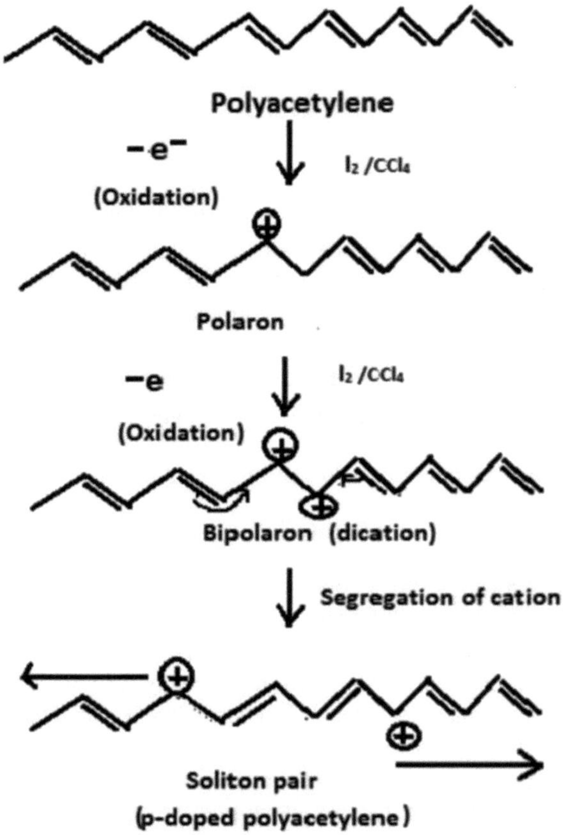

Conductivity of polymers made by above method is of the order 10–10 S cm^{-1}. Conductivity of polymers having only conjugated π-electrons is not enough for different applications. The polymer needs treatment to be able to be used in any applications: Doped Conducting polymers: polymers can be easily oxidized or reduced to make them intrinsically account using properties like low ionization energy and high electron affinities. The conductivity of these polymers can be increased by inducing charge on polymer background by oxidation or reduction.

Doping can be of two types (i) p-doping:—When the polymer is treated with a Lewis acid, its oxidation takes place and holes (positive charges) are created on the polymer backbone. Commonly used dopants are I_2, $FeCl_3$, Br_2, etc.

By exposing the conducting polymers to halogen vapors, the energy gap is reduced and the carrier density is increased to achieve a different level of semiconducting (or even metallic) behavior (Fig. 4.1).

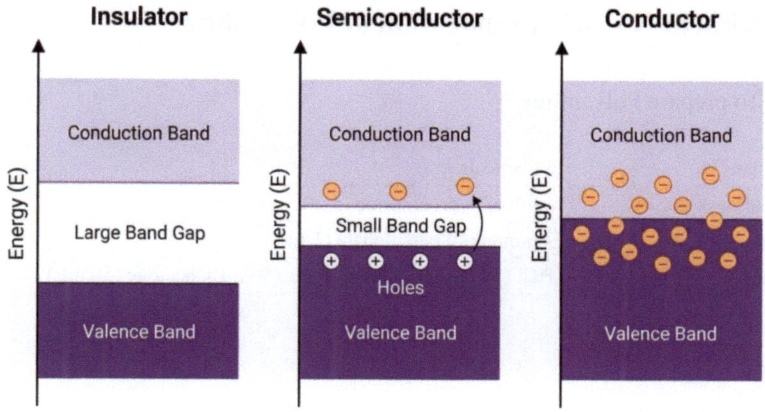

Fig. 4.1 Schematic evolution of the band structure moving from an insulator to a semiconductor to a conductor [1]

From the discovery of conducting polymers, organic chemists understood that the use of hetero-aromatic rings can enable the development of undoped polymers with intrinsic conducting properties. Thus, in 1980s some of the important Conducting Polymers such as polyaniline and polypyrrole were synthesized. Unlike inorganic semiconductors, in conducting polymers, the atoms are held strongly via covalent bonds. Conducting Polymers consist of macromolecules that are connected via secondary interactions like π–π, van der Waals and dipolar interactions. These forces are collectively strong enough to prevent their processing with solution-based methods.

Conducting Polymers applications largely depend on their:

a. processing characteristics,
b. the doping (charge density) level,
c. redox properties, and
d. whether the charge transport is purely electronic or mixed ionic/electronic.

The first Conducting Polymer [1] used was in areas where heavily doped/highly conducting materials are required. It was noticed that any object placed in the reaction vessel for the electrochemical synthesis of polyaniline or polypyrrole, get coated with a highly conducting polymer film. This opened the usage CPs as electrode materials in batteries and in antistatic/anticorrosion coating layers.

Recent studies revealed that Conducting Polymer-based lithium ion batteries perform far better than that of the inorganics.

Conducting Polymers can be synthesized by various electrochemical and chemical methodologies such as oxidative or reductive polymerization and polycondensation reactions. The overall cost of Conducting Polymers depends on several factors, like the cost of the monomer and reagent, the polymerization procedure, product isolation and purification steps, waste management, and location of production, etc.

Experiment No. 4.1: Preparation of Polyaniline

Aim: To prepare Polyaniline

Apparatus	Chemicals
Conical flask	Aniline hydrochloride (5 mL)
Magnetic stirrer	Ammonium persulphate (1 g)
Magnet	HCl (100 mL), distilled water (50 mL), acetone (50 mL)

Theory

Polyaniline (PANI) is a conducting polymer invented in the 1980s. This is a polymer made by polymerizing aniline. Due to the presence of conjugated double bonds, the labile pi electrons can carry the charge within its structure. PANI is doped by different acids and dopants as it is easy to synthesize and have high environmental stability. Polyaniline can be prepared by chemical oxidation method, electrochemical polymerization, enzyme-catalyzed polymerization, reversed micelle polymerization etc. PANI exists in three forms:

- Leucoemeraldine (LB): This is the fully reduced form of PANI, appearing colorless or white. It is a non-conducting insulator.
- Emeraldine (EB): This is the half-oxidized form, typically appearing blue in its base form and green in its salt form. It exhibits conductivity due to the presence of polarons and bipolarons.
- Pernigraniline (PB): This is the fully oxidized form, appearing blue to violet. It is also a conductive polymer.

It is used as sensors, drug delivery, anti-corrosion due to its unique properties, easy synthesis, low cost, and high environmental stability (Fig. 4.2).

Procedure

1. Oxidize 0.2 M aniline hydrochloride with 0.25 M ammonium peroxydisulfate in aqueous medium. Dissolve aniline hydrochloride 2.59 g in 50 mL of water in a flask.
2. Dissolve ammonium peroxysulfate (5.71 g) in 50 mL water.
3. Keep both the solutions at room temperature.
4. Then mix both the solutions in a beaker, stir and leave undisturbed to undergo polymerization.
5. After 24 h collect the precipitated aniline on a filter paper, wash with 0.2 M HCl three times followed by acetone.
6. Dry the polyaniline hydrochloride powder in air and then in vacuum at 60 °C.

Observation: Check the conductivity of the resultant polymer.

Result: The yield of the polyaniline prepared is ....%.

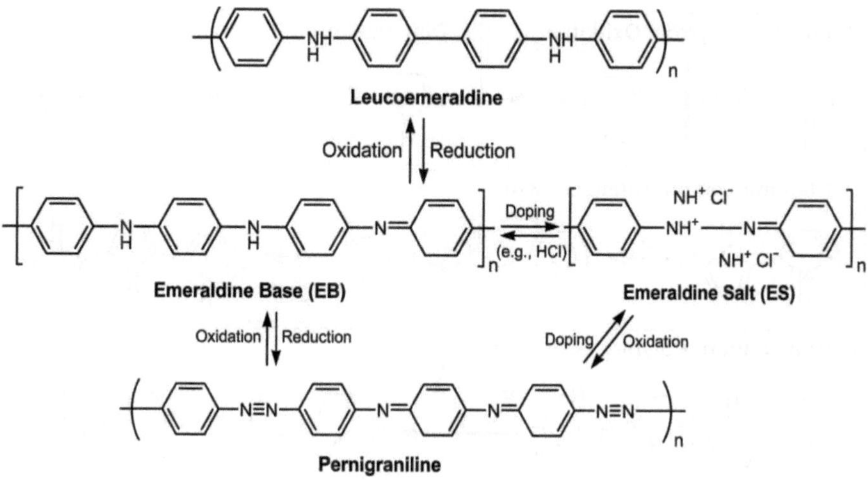

Fig. 4.2 All the three forms of PANI

The formation of polyaniline can be confirmed using FTIR and other identification tests mentioned in Chap. 1.

Precautions

1. Careful control of reaction conditions like pH, dopant selection, and monomer concentration is essential.

Experiment No. 4.2: Preparation of Polypyrrole

Aim: To prepare polypyrrole

Apparatus	Chemicals
Round bottom flask	Pyrrole (96%)
Magnetic stirrer	Sodium dodecyl benzene sulphonate
Magnet, ice bath, filter paper	Iron (III) chloride (99%) and HCl (37%)

Theory: Polypyrrole (PPy) is an organic polymer formed by the oxidative polymerization of pyrrole. It is a conducting polymer with a solid structure, and it's used in various applications like electronics, optics, and medicine. PPy's unique properties, such as high conductivity and environmental stability, make it a valuable material in diverse fields.

The biocompatible feature of polypyrrole accentuates its potential biomedical application. Due to its good chemical stability in air and water, polypyrrole is widely studied for implantable biomedical devices in preclinical studies (Scheme 4.1).

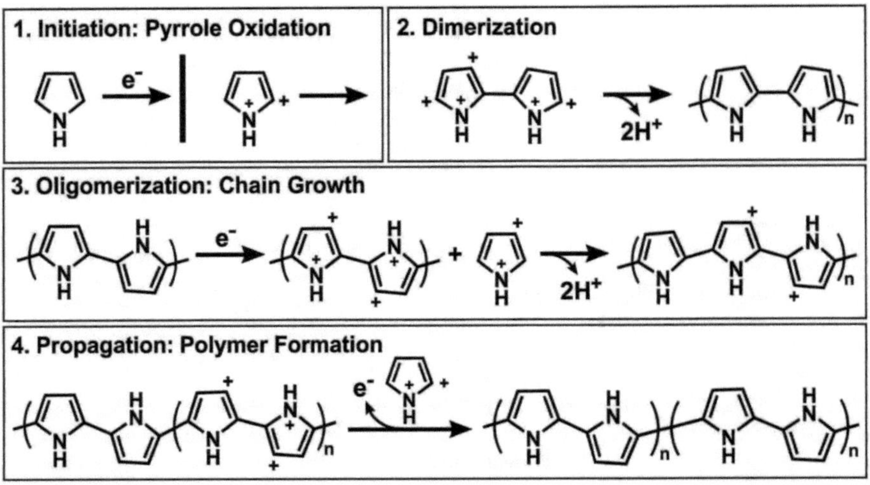

Scheme 4.1 Preparation of polypyrrole

Procedure

1. In a container take 3.48 g sodium dodecyl benzene sulfonate (surfactant) is dissolved in 100 mL of water.
2. In another beaker 27.03 g $FeCl_3{\cdot}6H_2O$ is dissolved in 100 mL of water.
3. Both the solutions are then mixed in a reaction flask and stirred for 15 min at 0 °C.
4. Now a solution of 0.15 mol pyrrole is added dropwise into the reaction mixture for different reaction times.
5. Now the precipitate was filtered and washed with distilled water several times.
6. The Black precipitate of polypyrrole was dried at 80 °C for 8 h.

Observation Table

Temperature of Reaction = 0 °C

Surfactant Used: sodium dodecyl benzene sulfonate (DBS-Na)

Sl. No.	Oxidant	Reaction time (min)	Conductivity	Yield (%)
1	$FeCl_3{\cdot}6H_2O$	20		
2	$FeCl_3{\cdot}6H_2O$	40		
3	$FeCl_3{\cdot}6H_2O$	60		
4	$FeCl_3{\cdot}6H_2O$	90		

Result: The amount of polypyrrole prepared is …g.

The formation of polypyrrole can be confirmed using FTIR and other identification tests mentioned in Chap. 1.

Precautions

1. Carefully control of reaction conditions, such as temperature, monomer concentration,
2. Control the type and concentration of dopant, to influence the polymer's structure and properties.

Experiment No. 4.3: Preparation of Polythiophene

Aim: To prepare Polythiophene

Apparatus	Chemicals
Beaker	Thiophene, chloroform
Magnetic stirrer	Anhydrous ferric chloride
Magnet	Dichloromethane
Filter paper	Acetonitrile

Theory

Polythiophene (PTh) can be synthesized through either electrochemical or chemical oxidation polymerization of thiophene monomers. Electrochemical methods involve applying a potential across a solution of the monomer, while chemical methods use oxidants like $FeCl_3$ or catalysts. Chemical synthesis offers advantages like greater monomer selection and potential for nanoparticle formation.

Polythiophene (PTh) and its derivatives are used in a wide range of applications due to their unique properties, including semiconductor qualities and electrical conductivity. These applications span from electronics and energy storage to biomedicine and environmental remediation

Procedure

1. Take 0.5 M thiophene and dissolve in 100 mL of dichloromethane in a round-bottom flask.
2. Freshly prepared anhydrous $FeCl_3$ solution is added to the thiophene solution with vigorous solution.
3. Formation of black precipitate confirms the formation of polythiophene.
4. The polymerization is allowed to proceed for 2 h with stirring.
5. The obtained precipitate is washed with acetone and acetonitrile till clear filtrate is obtained.
6. Now the precipitate is dried in an oven at 60 °C in a drying oven.

Result: Polythiophene is prepared and the yield is .....%.

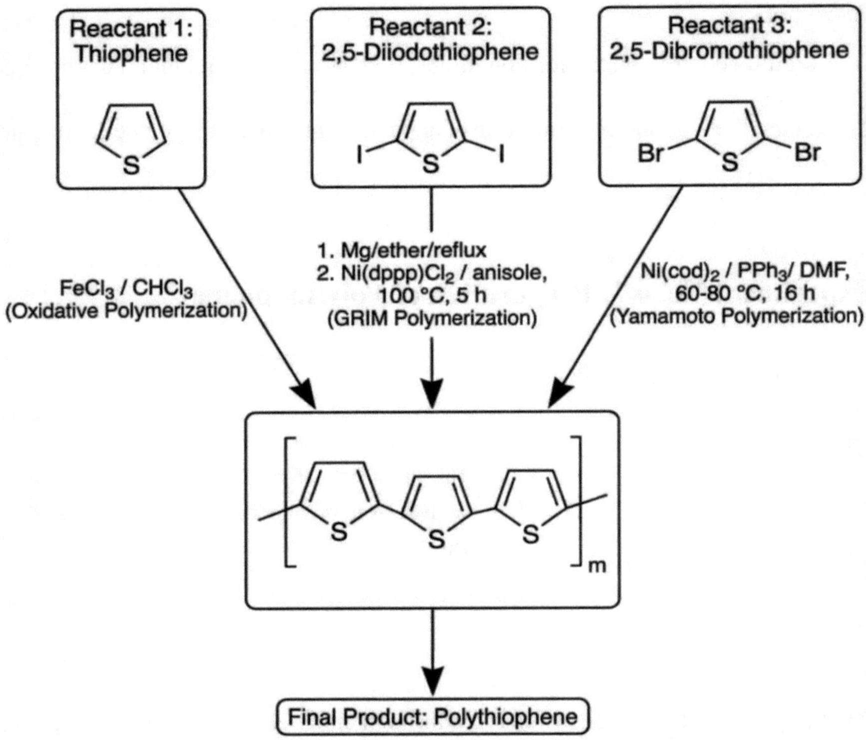

Scheme 4.2 Preparation of polythiophene

The formation of polythiophene can be confirmed using FTIR and other identification tests mentioned in Chap. 1.

Precautions

1. Carefully control the reaction conditions to avoid over-oxidation, selecting appropriate dopants for conductivity
2. manage the solubility issues associated with unsubstituted polythiophenes.

References

1. https://doi.org/10.1007/978-3-662-68921-9_23
2. Guo, X., & Facchetti, A. (2020). The journey of conducting polymers from discovery to application. Nature Materials, 19(9), 922–928. https://doi.org/10.1038/s41563-0200778-5

Chapter 5
Synthesis of Nanofillers

Abstract Nanofillers are inorganic or organic nanostructures which are added to the polymer matrix at a very level to enhance its various properties like mechanical property, thermal property, barrier property, and so on. These nanofillers are of different origin and different shapes and properties. Depending on the shape of the nanofiller the properties of the nanocomposite will vary. In this chapter preparation and properties of different types of nanofillers are discussed.

Keywords Nanofiller · Graphene · Nanoclay · Nanosilica · Nanocarbon

Fillers play a very important role in modification of polymers (both thermoplastic and thermosets). Fillers are used to reinforce tensile properties, thermal properties, barrier properties, optical properties, and so on. Some fillers are only used to reduce the cost and increase the bulk of the material. The basis of enhanced properties of filled composites is the interaction of the filler surface with the bulk of the polymer. This interaction is enhanced by the presence of functionality, surface area of the filler, and mixing quality.

Nanofiller is a special class of filler, where at least one dimension of the filler is in the nanometer level. These fillers owing to their small size, have a very high surface area available for interaction with the polymer chains. This leads to stronger bonding or interaction with the filler and enhances the properties. When it comes to mechanical properties the strong interaction of the nanofiller with the polymer matrix is responsible for enhancement in properties. This interaction will further increase if instead of only Vanderwal's interaction, H-bonding interaction, polar interaction, or ionic interaction is also present. That is why, in some of the nanoclays, the clay is modified with polar groups, so that the interaction of the nanoclay with the polar polymer matrix improves. Nanoclays based on montmorillonite are available in a whole range of polar to nonpolar. The clays are modified with a range of modifier starting from the one which is completely nonpolar to the one that has polar groups present in it. Thermal properties are also improved due to stronger interaction of the fillers with the polymer chains. These interactions or bonds require extra energy

S. D. Sadhu et al., *Source Book of Experimental Techniques and Analysis of Polymeric Materials*, https://doi.org/10.1007/978-981-92-2088-5_5

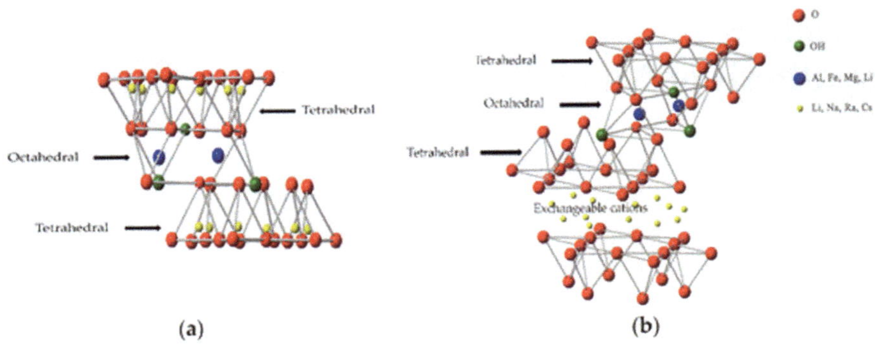

Fig. 5.1 Structure of **a** kaoline, **b** smectite

to undergo breakdown. This leads to the enhanced thermal stability and increased activation energy of the degradation of the nanocomposites.

Nanoclays are mainly organo-alumina-silicone in nature. A montmorillonite has a Tetrahedral-Octahedral-Tetrahedral (TOT) crystalline structure. The TOT structure is very stable and has a thickness of 1 nm. Between two such TOT layers there are Na ions present. These metal ions are weakly bonded and can be replaced with other cations by ion exchange reactions. Unlike montmorillonite, there are other types of nanoclays like bentonite, mica, etc. which also have different arrangements of tetrahedral and octahedral layers. The interlayer gap of the TOT layer has polar groups in the inner surface. This helped in the binding of the two layers by interlayer sodium ions. When these sodium cations are replaced with organic cations with bigger sizes, they push the two layers apart from each other. This way, when two TOT layers are pushed apart significantly, the layered structure breaks down and nanofillers are prepared. A composite of such nanofillers in a polymeric matrix results in the formation of nanocomposites (Fig. 5.1).

Due to the platelet structure, these nanofillers are oriented in the polymer matrix like fine sheets. Owing to their structure, they provide good resistance to solvents and fluids, enhancing the barrier property significantly. Other than that, the stronger interaction between the matrix and filler results in better mechanical, thermal, and physical properties. The nanometric particle does not hinder the passage of the light. This makes the nanocomposite more transparent in nature.

Metal nanofillers or metal oxide nanofillers are another category of nanofillers which also have their dimensions in the nanometer range. Mostly these nanofillers are globular or remain as agglomerates. Owing to their smaller size and higher surface area, metal-based nanofillers interact with the matrix better. This again leads to enhanced mechanical, thermal, and optical properties. Barrier properties are not as good as the nanoclays but improved compared to the conventional fillers. Metal/ metal oxides are known for their active nature. Metals/metal oxides are known as important micronutrients and they also play an important role in metabolic activities. This causes enhanced antibacterial property, antifungal behavior, and better antioxidant properties of metal/metal oxide-based nanocomposites. There are a range of metal-based nanocomposites which are commercially used in today's life. The names include nano silver, nano gold, nano zinc, nano copper, nano iron, and many more. One of the important aspects of nanometals is that due to the nanometric dimensions, these nanoparticles can enter the organs of a living being very easily. If not used carefully, these nanofillers can cause biohazards.

The third category of nanofillers are carbon nanotubes. These are graphene sheets rolled in the form of a tube. These are allotropes of carbon. These tubes may be of single-walled (approx. diameter 0.5–2.0 nm) or multiwalled where one tube is inserted inside another. Carbon nanotubes exhibit exceptional mechanical strength, rigidity, thermal conductivity, and electrical conductivity. In fact, in terms of strength and modulus, nanotubes are the strongest. This strength stems from the covalent sp^2 bonds formed between the individual carbon atoms. Some of the single-walled carbon nanotubes also behave as semiconductors. These properties are valuable in areas of electronics, composite materials, optics, nanotechnology, medicinal applications, and other fields of materials science.

Despite having extremely high strength of individual CNT shells, the weak shear interactions between two adjacent shells and tubes result in significant reduction in the strength of MWCNT. High energy irradiation may help in overcoming this issue by crosslinking the inner tubes. Due to high aspect ratio and hollow structure, they exhibit buckling under compressive, torsional, or bending stress.

Due to the presence of π-electron, the nanotubes exhibit superb conductivity. But doping in carbon nanotubes is different from that of crystalline semiconductors like the elements from the same group of the periodic table (e.g., silicon).

Carbon nanotubes (CNTs) are prepared using laser ablation, arc discharge, and chemical vapor deposition (CVD) methods. Each method is used to prepare nanotubes with a range of structures and properties that impact their applications. Some of the synthesis methods are as follows:

1. Arc Discharge: This method uses the passage of high-voltage electric arc between two graphite electrodes. It vaporizes carbon and forms nanotubes.
2. Laser Ablation: A high-powered laser is used to irradiate a graphite target, creating a carbon vapor plume that condenses into nanotubes.
3. Chemical Vapor Deposition (CVD): This method involves passing hydrocarbon gases through a catalyst reactor at high temperatures, causing carbon to deposit and form nanotubes on a substrate.

1. Flame Synthesis

CNTs can be produced by passing carbon-containing gases through a controlled flame, offering the potential for large-scale, cost-effective production.

2. Oxidation and Purification

After synthesis, CNTs may require purification to remove impurities and unwanted materials. This can involve oxidation techniques, such as treating the nanotubes with nitric acid (Fig. 5.2).

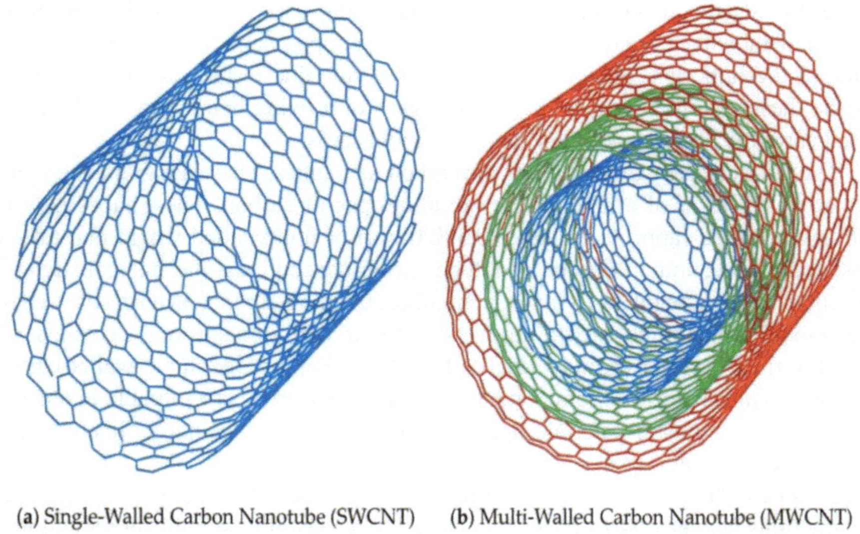

Fig. 5.2 **a** Single-walled carbon nanotube, **b** multi-walled carbon nanotube. Graphene comes next in the list of nanofillers. Graphene is a single layer of graphite where all the carbon atoms are sp^2-hybridized and form six-membered rings attached with each other. It has a two-dimensional structure with 1 nm thickness. Graphene has extremely high aspect ratio, high electrical and thermal conductivity due to labile pi-electron, high strength, and thermal stability

Graphene possesses exceptional properties including high strength, flexibility, electrical and thermal conductivity, and transparency. It is the world's thinnest material, only one atom thick. Graphene's unique properties make it suitable for a wide range of applications, including electronics, energy storage, and biomedical devices (Fig. 5.3).

Key Properties

- Strength: Graphene is 200 times stronger than steel but six times lighter.
- Flexibility: It is highly flexible and stretchable.
- Conductivity: It is an excellent conductor of both electricity and heat.
- Transparency: Graphene is highly transparent, absorbing only about 2% of visible light.
- Thickness: Graphene is the world's thinnest material, only one atom thick.
- Impermeability: It is impermeable to gases, even hydrogen and helium.
- Surface Area: Graphene has a very high surface area, which is beneficial for applications like energy storage and sensing.

Graphene is prepared through a number of methods:

1. "top-down" (exfoliation of graphite) and
2. "bottom-up" (growth of graphene).

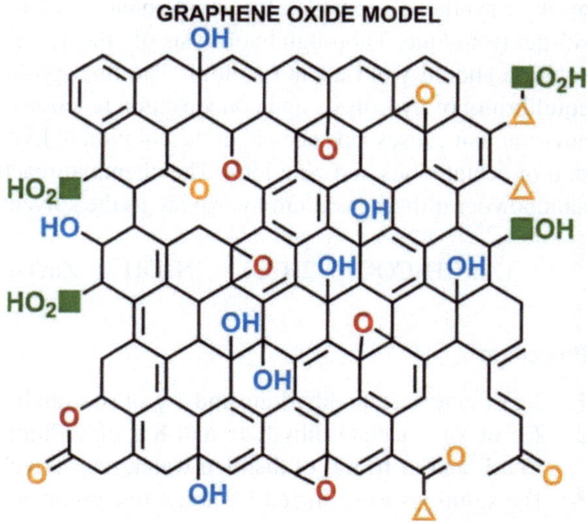

Fig. 5.3 Structure of graphene

Common methods of preparation include:

1. chemical vapor deposition (CVD),
2. mechanical exfoliation, and
3. chemical exfoliation (oxidation of graphite to graphene oxide followed by reduction).

Experiment No. 5.1: Synthesis of ZnO Nanoparticles

Aim: Synthesis of ZnO nanoparticles

Apparatus	Chemicals
Conical flask	Zinc acetate dihydrate ($Zn(CH_3COO)_2 \cdot 2H_2O$) with ≥99% purity
Magnetic stirrer	Sodium hydroxide (NaOH) with ≥98% purity
Magnet	Ethanol (CH_3COOH)

Theory: Metal-based nanoparticles have drawn a lot of attention due to their distinct characteristics. For synthesizing metal-based nanomaterials, various methods have been proposed including technologies like chemical vapor condensation, arc discharge, hydrogen plasma-metal reaction, laser pyrolysis in the vapor phase, microemulsion, hydrothermal, sol–gel, sonochemical, and microbial processes in the liquid phase, as well as ball milling in the solid phase. The particle characteristics like size, color, shape etc., of the metal nanoparticles are predominantly influenced by their synthesis methods. Zinc oxide nanostructures were produced utilizing the sol-gel technique. Thorough hydrolysis of zinc acetate using NaOH in an ethanolic solution should yield a ZnO colloid. The final product was achieved through the equilibrium of hydrolysis and condensation reactions. The heating in a hydrolyzing environment causes zinc acetate in the solution to hydrolyze, resulting in the formation of acetate ions and zinc ions. The chemical reaction for the formation of ZnO nanopowder utilizing sodium hydroxide as the solvent is as follows

$$(Zn(CH_3COO)_2 \cdot 2H_2O) + 2NaOH \rightarrow ZnO + 2NaCH_3COO + H_2O$$

Procedure

1. 2 g of zinc acetate dihydrate and 8 g of sodium hydroxide was weighed.
2. 2 g of zinc acetate dihydrate and 8 g of sodium hydroxide were dissolved in 15 mL and in 10 mL of distilled water, respectively.
3. The solutions were stirred for about five minutes each.
4. First the sodium hydroxide solution was poured into the zinc acetate solution with constant stirring for about five minutes.
5. Now, ethanol is added dropwise to the solution of sodium hydroxide and zinc acetate solution to slowly obtain a white precipitate on ZnO.
6. After filtration of the ZnO nanofiller, it was washed with alcohol and dried.

Observation

Initial wt of zinc acetate	Wt of ZnO	Percentage Yield

Result

ZnO nanoparticle obtained after the reaction is .... g.

The formation of ZnO nanoparticles can be confirmed using FTIR and Scanning Electron Microscopy (SEM).

Precautions

1. Use appropriate protective equipment,
2. handle hazardous chemicals carefully,
3. manage waste responsibly,
4. consider the potential for toxicity and environmental impact.

Experiment No. 5.2: Preparation of Copper Nanoparticle

Aim: Preparation of copper(I) Cu_2O and Copper(II) CuO nanoparticle.

Apparatus	Chemicals
Conical flask	NaOH
Magnetic bead and magnetic stirrer	Fructose/glucose
Burner, funnel, Whitman filter paper	$CuSO_4·5H_2O$
Dropper, aluminium foil	Water

Theory: Copper-based nanoparticles are available in less than 100 nm range. Copper-based nanoparticles used to be prepared as glazed glasses and ceramics in ninth century in Mesopotamia. Copper is available in two oxidation states (+1 and +2). Thereby copper oxides can be prepared in two forms Cu_2O and CuO. Copper plays a very important role as a micronutrient. It helps in digestion in the gut.

Copper nanoparticles with a high surface-to-volume ratio can serve as anti-fungal/antibacterial agents. These nanoparticles interact very closely with microbial membranes, inducing their antimicrobial activity. On slow oxidation of these nanoparticles in solutions, cupric ions create toxic hydroxyl free radicals when the lipid membrane is nearby.

Copper nanoparticles having great catalytic activities find applications in biosensors and electrochemical sensors.

Procedure

A. Preparation of Cu_2O

1. Weight 4 g of $CuSO_4·5H_2O$ and 2.4 g of fructose.
2. Take 40 mL of water in a conical flask and put it on the magnetic stirrer.
3. Add $CuSO_4·5H_2O$\slowly.
4. When the $CuSO_4·5H_2O$ has dissolved then slightly heat the solution (on burner).
5. Now place the conical flask on the stirrer and add fructose slowly with continuous stirring.
6. Then, again slightly heat the mixture on the burner.
7. Now prepare 20% solution of NaOH by mixing of 5 g NaOH in 25 mL of water.
8. Now add 20% NaOH solution drop by drop in the solution of $CuSO_4$ with continuous stirring and slight heating on the burner.
9. Continue to add 20% NaOH solution till the blue color is not converted to yellow color.
10. Now allow the precipitate to settle down without disturbance for 1–2 h.
11. Now filter the precipitate and wash with the water.
12. After washing, dry the precipitation at 65 °C for 3–4 h and then report the yield.

B. Preparation of CuO

1. Weight 4 g of $CuSO_4·5H_2O$ and 10 g NaOH seperately.
2. In a conical flask add $CuSO_4·5H_2O$ in 40 mL of water.

3. Stir the mixture with a magnetic stirrer.
4. When $CuSO_4 \cdot 5H_2O$ is dissolved completely, then heat the solution slightly.
5. Now, prepare 20% solution of NaOH by mixing 5 g NaOH in 50 mL water.
6. Add 20% NaOH solution drop by drop in the solution of $CuSO_4$ with continuous stirring and slightly heat the mixture.
7. Continuously add 20% NaOH solution till the blue color is converted into a brownish-black color.
8. Now allow the precipitate to settle down without disturbance for 1–2 h.
9. Filter the precipitation and wash with water.
10. After washing, dry the precipitation at 65 °C for 3–4 h and then report the yield.

Result

The percentage yield of Cu_2O and CuO are .... and .... respectively.

The formation of CuO and Cu_2O nanoparticles can be confirmed using FTIR and Scanning Electron Microscopy (SEM).

Precautions

1. Use appropriate protective equipment,
2. handle hazardous chemicals carefully,
3. manage waste responsibly,
4. considering the potential for toxicity and environmental impact.

Experiment No. 5.3: Preparation of Silica Nanoparticle

Aim: To prepare nanosilica by sol gel method

Apparatus	Chemical
Beaker	TEOS
Dropper, magnetic stirrer	Ethanol
Filter paper, funnel and glass rod	Ammoniated water, deionized water

Theory: Silica is a well-known reinforcing filler which is used for a long time for making polymer composites. Silica has a lot of active hydroxyl groups on its surface. With the formation of nanosilica, the size of the silica particles is reduced to nanolevel. Formation of nanosilica results in a huge increase in its surface area which in turn is responsible for high surface energy.

Procedure

1. Choose a suitable silicon precursor (Tetraethyl Orthosilicon, TEOS), which is readily available and provides good control over the reaction.
2. Dissolve the silicon precursor in a solvent like ethanol in a beaker and adjust the concentration (10%) to achieve the desired particle size.
3. Add a catalyst, either a dilute acid (e.g., HCl or base like ammonia in deionized water) to initiate the hydrolysis and condensation reactions. The mixing of acid/ base in TEOS solution should continue with constant stirring at room temperature for 90 min.
4. The silicon alkoxide bonds are broken by water molecules, forming silanol groups (Si–OH). The silanol groups react with each other, forming siloxane bonds (Si–O–Si) and creating a growing network of silica particles.
5. As the network continues to grow, the solution becomes viscous and eventually forms a gel.
6. The gel is typically allowed to age for a period (24 h) to further develop the network structure and control particle size.
7. The gel is then washed with deionized water to remove impurities till the pH is neutral, followed by drying at 100 °C for 4 h to obtain the nanosilica powder.

Observation

The SiO_2 precipitate obtained is white in color Result:

The amount of SiO_2 obtained is .... g.

The formation of SiO_2 nanoparticles can be confirmed using FTIR and Scanning Electron Microscopy (SEM).

Precautions

1. Use appropriate protective equipment,
2. handle hazardous chemicals carefully,
3. manage waste responsibly,
4. considering the potential for toxicity and environmental impact.

Experiment No. 5.4: Preparation of Nanoclay

Aim: Preparation of organically modified nanoclay

Apparatus	Chemicals
Beaker	Montmorillonite clay
Magnetic stirrer	Dodecyl amine, HCl
Magnet bead, filter paper	Water

Theory

Clays are used as fillers in polymer composites for a long period of time. Normally clay is known to be a non-reinforcing filler. But modification of the clay with organic hydrocarbons turns the natural clay into nanofiller with enhanced properties. Montmorillonite, a clay from the smectite group of clay has a tetrahedral-octahedral-tetrahedral sheet-like structure. Once the interlayer gap is intercalated with organic groups, the interlayer gap increases.

Beyond a certain limit, the layered orientation breaks down and nanofillers are formed (Fig. 5.4).

Procedure

Choose a clay mineral, most commonly montmorillonite (MMT) due to its high cation exchange capacity and layered structure.

- Grind the raw clay into a fine powder.
- Suspend the clay in water. Stir the suspension to achieve a homogeneous mixture using a magnetic stirrer for 30 min.
- Now take 2 g of dodecyl amine in a beaker and add conc. HCl to it with stirring with glass rod. Now immediately add the ammonium salt into the clay suspension.
- The organic molecules will interact with the clay surface through electrostatic attraction or chemical bonding. Let the reaction continue for 30 min with stirring.
- Now allow the modified clay to settle down.
- Filter the precipitated clay and wash it followed by drying.

Result

The dried nanoclay prepared was weighed. The nanoclay obtained is ….. g.

The formation of nanoclay nanoparticles can be confirmed using FTIR and Scanning Electron Microscopy (SEM).

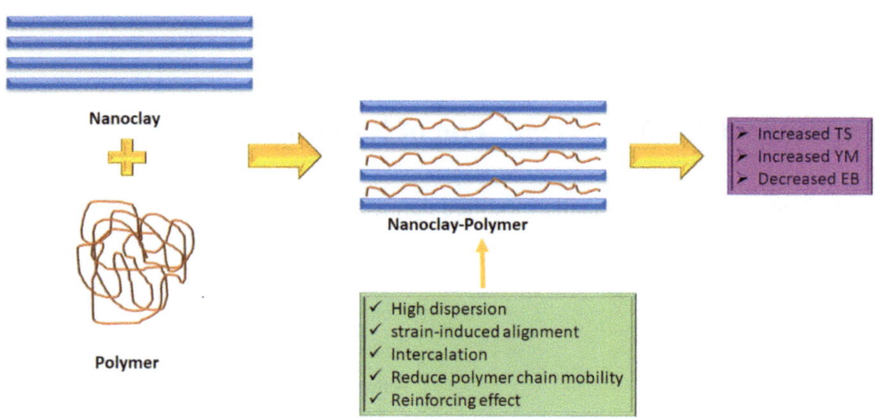

Fig. 5.4 Schematic diagram of nanoclay formation [https://doi.org/10.1007/s40097-023-00525-5]

Precautions

1. Use appropriate protective equipment,
2. handle hazardous chemicals carefully,
3. manage waste responsibly,
4. considering the potential for toxicity and environmental impact.

Experiment No. 5.5: Preparation of Graphene

Aim: Preparation of graphene oxide from graphene

Apparatus	Chemicals
Beaker	Graphite powder, distilled water, ice bath
Magnetic stirrer, magnetic bead	Potassium powder
Dropper, filter paper, funnel	Sulfuric acid, hydrogen peroxide

Theory

This approach can create high-quality graphene with a minimal number of flaws. Researchers have documented the one-step synthesis of high-quality graphene oxide (GO) and reduced graphene oxide (rGO) using chemical exfoliation, utilizing H_2SO_4/ $KMnO_4$ in conjunction with ultrasonication for the preparation of rGO. By regulating the reaction parameters, they successfully synthesized GO and rGO with a minimal defect count. Chemical exfoliation is an effective synthesis method; nevertheless, it typically employs highly harmful oxidizing chemicals like $KMnO_4$, rendering it less environmentally sustainable. Focus has shifted to employing less toxic compounds as oxidation agents, with numerous experiments undertaken on chemical exfoliation utilizing these safer alternatives. A recent study employed nitric acid (HNO_3) and sulfuric acid (H_2SO_4) to initiate the exfoliation of graphite, subsequently utilizing an ammonium hydroxide (NH_4OH) solution to finalize the exfoliation and partially decrease the resulting graphene samples (Fig. 5.5).

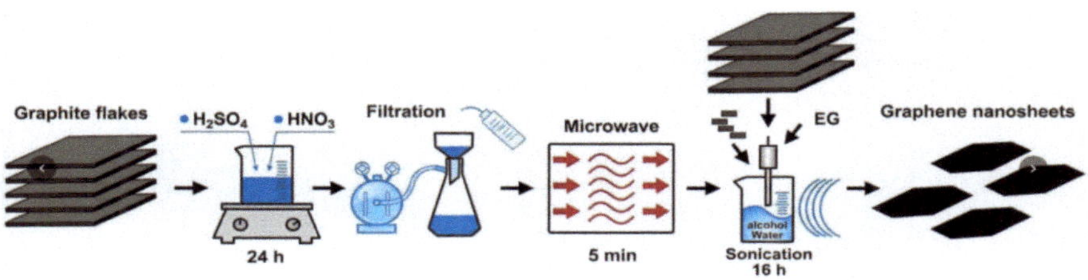

Fig. 5.5 Schematic diagram of graphene formation

Procedure

1. Take a magnetic stirrer and set an ice bath.
2. Add 25 mL sulfuric acid.
3. Add 1 g graphite powder.
4. Add 3 g potassium permanganate slowly for 3 h.
5. Add 50 mL water dropwise. Extra care needs to be taken as the reaction is extremely exothermic and explosive.
6. Now add 10 mL of water instantaneously.
7. Add 5 mL of hydrogen peroxide.
8. Graphene oxide thus formed is collected at the bottom.
9. The graphene oxide thus obtained is washed with water till the filtrate is neutral in pH.

Observation

The amount of graphene oxide is .... g.

The formation of graphene nanoparticles can be confirmed using FTIR and Scanning Electron Microscopy (SEM).

Precautions

1. Use appropriate protective equipment,
2. handle hazardous chemicals carefully,
3. manage waste responsibly,
4. considering the potential for toxicity and environmental impact.

Appendix A
Structure and Properties of Polymers

Polyethylene (PE)

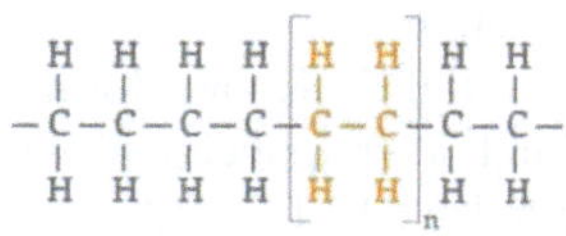

Polyethylene (PE) is a widely used thermoplastic polymer, known for its versatility and affordability. It's a hydrocarbon polymer, primarily used in packaging applications, and is produced in various densities, including low-density polyethylene (LDPE), linear low density polyethylene (LLDPE), and high-density polyethylene (HDPE).

Key Properties

- Chemical Resistance: PE is highly resistant to many chemicals, including alcohols, dilute acids, and alkalis.
- Flexibility and Impact Resistance: It's known for its flexibility and ability to withstand impacts, making it suitable for packaging and containers.
- Electrical Insulation: PE is a good insulator, making it useful for electrical wire and cable sheathing.
- Low Water Absorption: PE absorbs very little water, making it useful in applications where moisture resistance is important.
- Low Cost and Easy Processing: It's relatively inexpensive and easy to process using various methods like injection molding and extrusion.
- Low Friction: PE has a low coefficient of friction, which makes it useful in applications where sliding or moving parts are involved.

S. D. Sadhu et al., *Source Book of Experimental Techniques and Analysis of Polymeric Materials*, https://doi.org/10.1007/978-981-92-2088-5

Important Applications

- Packaging: PE is widely used in films for packaging food, groceries, and other products, as well as in bottles, containers, and bags.
- Construction: PE pipes and fittings are used in plumbing, drainage, and irrigation systems. PE sheets are used for various applications, including construction and agriculture.
- Electrical and Electronics: PE is used for insulation of wires and cables, and in some electronic components.
- Automotive: PE is used in automotive parts, such as fuel tanks and interior components.
- Medical: PE is used in medical devices, such as artificial joints and drug delivery systems.
- Other: PE is also used in toys, household goods, and sporting equipment.

Types of Polyethylene

- Low-Density Polyethylene (LDPE): Known for its flexibility and is commonly used in films, bags, and packaging.
- High-Density Polyethylene (HDPE): Known for its strength and rigidity, used in bottles, containers, pipes, and other applications where durability is important.
- Linear Low-Density Polyethylene (LLDPE): A copolymer of ethylene and another alpha-olefin, offering a good balance of flexibility and strength.
- Ultra-High Molecular Weight Polyethylene (UHMWPE): Extremely strong and wear-resistant, used in specialized applications like bearings and joint replacements.

Polypropylene (PP)

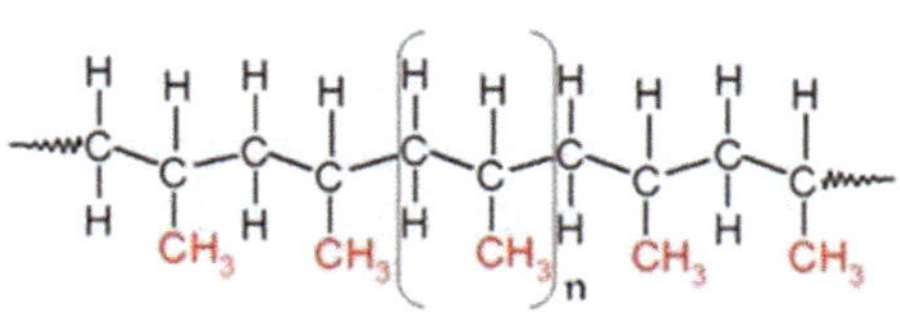

Polypropylene (PP) is a thermoplastic polymer extensively utilized for its characteristics and cost-effectiveness.

Key Properties

- Polypropylene is lightweight, rendering it appropriate for applications where weight is a critical factor.
- It exhibits resistance to a broad spectrum of chemicals, encompassing several organic solvents, acids, and alkalis.

- Polypropylene (PP) can endure mild temperatures, rendering it appropriate for uses such as food packaging and microwave-safe containers.
- It withstands stress cracking and flexural deformation, rendering it appropriate for applications such as hinges.
- PP exhibits considerable toughness and may undergo deformation without fracturing in the initial stages, enhancing its elasticity.
- Although typically opaque, polypropylene can be manufactured as a transparent material based on the use.
- PP serves as an effective electrical insulator, rendering it appropriate for applications necessitating insulation.
- It has significant resistance to moisture absorption, rendering it appropriate for liquid packaging.

Important Applications

Polypropylene (PP) is extensively utilized for packaging owing to its robustness, moisture resistance, and ease of sealing.

- Illustrations encompass:
  - Food containers and packaging: Microwaveable containers, yogurt cups, hot beverage cups, and food packaging materials.
  - Packaging for diverse products: apparel, tobacco items, and additional merchandise.
- Rigid packaging includes pallets, crates, bottles, jars, and thin-walled containers.
- The automotive industry utilizes polypropylene in components such as bumpers, interior elements, and cladding. It is additionally utilized in battery enclosures.
- Textiles: Polypropylene fibers are utilized in carpets, furniture, apparel, and camping gear.
- Medical Devices and Applications: Polypropylene (PP) exhibits biocompatibility and can undergo sterilization, rendering it appropriate for medical devices, syringes, intravenous components, and medical packaging.

Alternative Applications

- Electrical and electronics: Wiring harnesses, connectors, and additional components owing to their insulating qualities.
- Household and consumer products: Storage containers, toys, kitchenware, and outdoor furniture.
- Industrial applications include pipes, tanks, chemical containers, industrial filters, and battery enclosures.
- Material handling: Pallets and various material handling apparatus.
- Membrane technology: Improves filtration and antifouling characteristics.
- Water remediation: Utilized as a sorbent in the treatment of wastewater.

Polystyrene

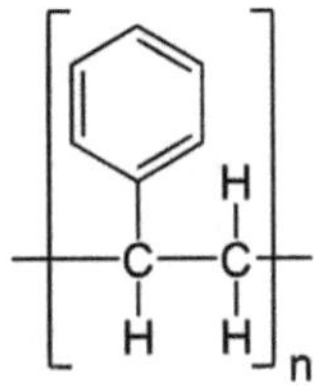

Polystyrene is a synthetic polymer derived from the monomer styrene. The bulk polymerization process might be employed for its preparation. Polystyrene is a clear, inflexible thermoplastic polymer. The density is around 1.05 g/cc. It is utilized in applications such as CDs, packaging, food trays, foams, and the electronics sector.

Key Properties

- Transparency: Inherently clear and capable of being tinted.
- Rigidity and brittleness: Firm and inflexible, yet susceptible to fracture under stress.
- Electrical Insulator: Suitable for situations necessitating insulation.
- Chemical Resistance: Resistant to acids and bases, but comparatively less resistant to solvents and ultraviolet light.
- Economical: Comparatively affordable to manufacture and utilize.
- Lightweight: Convenient to manage and transport, especially when foamed.
- Thermal Insulation: Foam variants provide effective thermal insulation.
- Dimensional Stability: Preserves its form and dimensions effectively.
- Moisture Resistance: Impervious to moisture absorption.
- Inadequate Impact Resistance: May exhibit brittleness and susceptibility to fracturing.
- Inadequate Oxygen and UV Resistance: Vulnerable to deterioration from oxygen and ultraviolet radiation.

Important Applications

- Packaging:
  - Polystyrene is utilized in the production of disposable cups, plates, bowls, and various food containers, particularly for hot and cold beverages and food items.
  - Protective packing serves as cushioning for fragile objects during transportation and storage, exemplified by packaging peanuts and foam inserts.
  - Film packaging: Delicate polystyrene sheets are employed to encase or shield things for safeguarding and presentation.
- Construction:

  - Insulation: Polystyrene foam, encompassing expanded polystyrene (EPS) and extruded polystyrene (XPS), is a widely utilized insulation material for walls, roofs, and floors, offering both thermal and acoustic insulation properties.
  - Construction materials: They might be utilized as sheets, panels, or blocks for diverse construction purposes, encompassing insulation, cladding, and lightweight construction.

- Electronics and Appliances:
  - Polystyrene is utilized for enclosures, panels, and various components in electronics, appliances, and other equipment.
  - Sound attenuation: It can be utilized to diminish noise in electrical devices and appliances.

- Additional Sectors:
  - Polystyrene is utilized in the manufacturing of toys and various plastic products.

- Biomedical applications:
  - Polystyrene is utilized in the production of medical devices, implants, disposables, and laboratory equipment such as Petri dishes and pipettes.
  - Polystyrene is utilized in the production of disposable cutlery, plates, and various food service products.
  - Transportation: It is applicable in the fabrication of automobile components, including knobs and instrument panels.

Categories of Polystyrene

- General Purpose Polystyrene (GPPS): Utilized manufacturing disposable cups, CD covers, and various transparent items.
- High Impact Polystyrene (HIPS): More resilient and impact-resistant, utilized in appliances, electronics, and packaging.
- Expanded Polystyrene (EPS): A lightweight foam utilized for insulation, packaging, and many purposes.
- Extruded Polystyrene (XPS): A variant of foam frequently utilized for insulation purposes.

Polycarbonate

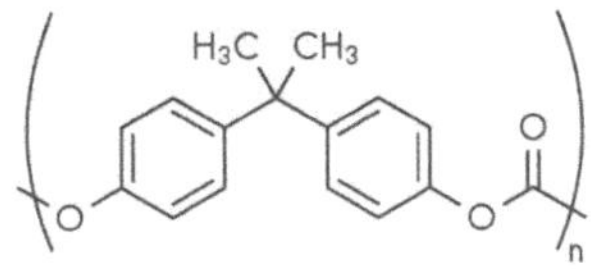

polycarbonate

Polycarbonate (PC) is a clear, durable polymer recognized for its exceptional impact resistance, commendable chemical resistance, and thermal stability. These characteristics render it appropriate for diverse applications, encompassing automotive, construction, electronics, and medical devices.

Key Properties

- Superior Impact Resistance: Polycarbonate (PC) exhibits markedly greater strength than glass, rendering it optimal for applications necessitating durability and safety.
- Transparency: Polycarbonate may be rendered translucent, rendering it appropriate for applications such as windows, skylights, and lenses.
- Thermal Resistance: Polycarbonate (PC) can endure a broad spectrum of temperatures, rendering it appropriate for applications involving heat exposure.
- Chemical Resistance: Polycarbonate exhibits resistance to numerous prevalent compounds, encompassing acids, bases, and alcohols.
- Lightweight: Notwithstanding its durability, polycarbonate is lightweight, facilitating ease of handling and transportation.
- Dimensional Stability: Polycarbonate retains its form and dimensions effectively, even when subjected to stress or fluctuations in temperature.
- Low Flammability: Polycarbonate has low flammability, rendering it appropriate for applications where fire safety is paramount.

Important Applications

- Automotive: Headlamps, car glazing, interior elements, and additional components.
- Construction: Roofing, skylights, safety glazing, and machinery guards.
- Electronics: Laptop and smartphone enclosures, electrical components, and telecommunications apparatus.
- Eyewear: Protective glasses, athletic goggles, and additional optical devices.
- Medical Devices: Surgical instruments, diagnostic apparatus, and further medical equipment.
- Additional Applications: Compact discs, infant feeding bottles, protective helmets, ballistic glass, among others.

Silicone

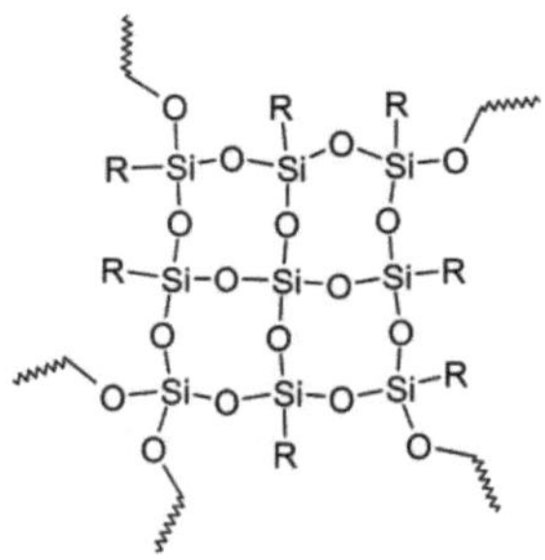

Silicone is a multifaceted substance recognized for its pliability, thermal resilience, low toxicity, and hydrophobic properties. It is utilized in several applications, including medical devices, cookware, and industrial sealants. Essential characteristics encompass electrical insulation, chemical stability, and resilience to severe temperatures.

Key Properties

- Suppleness and Durability: Silicone may be shaped into diverse forms and maintains its suppleness even under high temperatures.
- Thermal Resistance: Silicone endures a broad spectrum of temperatures, rendering it appropriate for uses such as culinary utensils and electrical insulation.
- Water Repellency: Silicone exhibits hydrophobic properties and can create impermeable seals.
- Minimal Toxicity: Silicone is widely regarded as safe for culinary and medicinal purposes, having received FDA approval for such use.
- Electrical Insulation: Silicone possesses superior electrical insulation characteristics, rendering it appropriate for electrical apparatus and cabling.
- Chemical Stability: Silicone exhibits resistance to oxidation, ozone, and ultraviolet radiation, rendering it durable in adverse conditions.

Important Applications

- Medical Devices: Silicone is utilized in implants, medical apparatus, and respiratory masks owing to its biocompatibility.
- Cookware and Food Storage: Silicone is utilized in cookware, utensils, and food storage containers because of its thermal resistance and non-stick characteristics.
- Industrial Applications: Silicone is utilized in O-rings, seals, gaskets, caulks, and flexible molds throughout diverse industries.
- Electrical Insulation: Silicone is utilized in transformers, voltage line insulators, and various electrical apparatus.
- Cosmetics: Silicone serves as a smoothing and conditioning agent in hair conditioners, shampoos, and cosmetics.

- Waterproofing Systems: Silicone is utilized in waterproofing applications including bathrooms, roofs, and piping.
- Silicone rubber is utilized in aerospace and defence applications for components such as gaskets and seals owing to its thermal resilience.

Polymethylmethacrylate

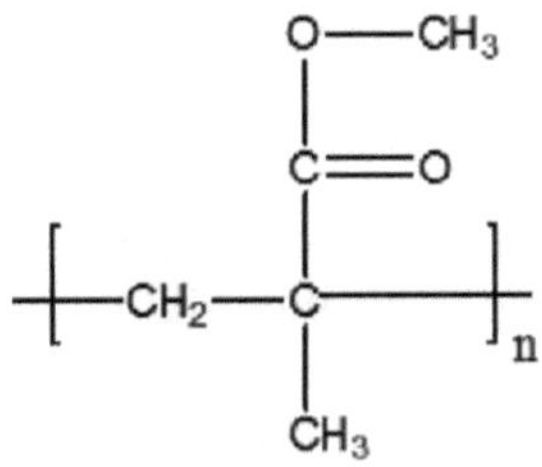

PMMA is a thermoplastic belonging to the acrylic polymer group, characterized by its rigidity, transparency, high strength, and UV resistance. It possesses optical clarity, elevated impact resistance, and superior weatherability. PMMA possesses an amorphous structure, which enhances its transparency. It is composed of the polymerization of methyl methacrylate monomers. It is utilized in signage, optical lenses, and safety glass applications.

Key Properties

- Superlative optical clarity (exceeding that of glass) facilitating elevated light transmission.
- Elevated strength, hardness, and impact resistance.
- Exhibits excellent thermal stability and resistance to temperature variations.
- Exhibits significant resistance to UV radiation, rendering it appropriate for outdoor applications.
- Soluble in organic solvents.
- Comparatively lightweight, possessing a density of approximately 1.2 g/cm^3.

Important Applications

Construction and Architecture

- Windows and Skylights: PMMA serves as a favored substitute for glass due to its robustness and resilience to impact.
- Aquariums: Its transparency and shatter resistance render it optimal for substantial aquarium tanks.
- Signage and Displays: PMMA may be readily molded and tinted, rendering it appropriate for many signage and display uses.
- Transparent PMMA roofing panels are applicable in greenhouses and sports complexes.

Automobile Sector

- Exterior Illumination: PMMA is utilized for lenses in headlights, taillights, and several other exterior lighting elements.
- Trim and Panels: It is available in both interior and exterior applications.
- Windows: Certain automotive windows and windshields are constructed with PMMA.

Applications in Medicine and Dentistry

- Dental Prosthetics: PMMA is utilized in the production of artificial teeth, dentures, and various dental prostheses.
- Bone Cement: PMMA serves as a bone cement in orthopedic procedures for joint replacement and bone stabilization.
- Medical Devices: It is applicable in incubators, pharmaceutical testing apparatus, and storage units within hospitals and laboratories.
- Intraocular Lenses: PMMA is utilized in the production of intraocular lenses.

Styrene and Methylmethacrylate Copolymer

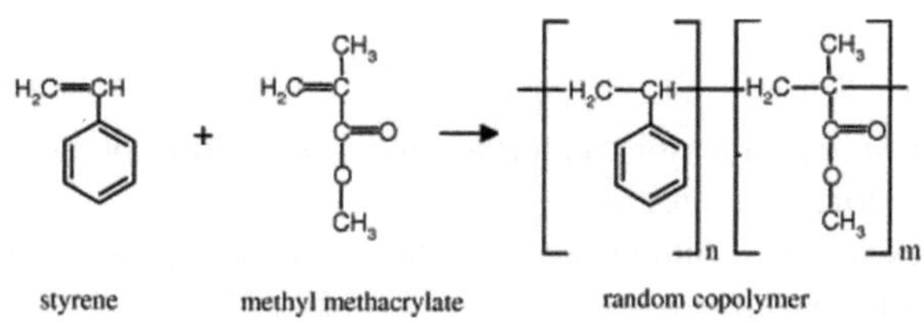

Styrene-methyl methacrylate (S-MMA) copolymers display various qualities contingent upon their composition. These copolymers may exhibit transparency, rigidity, and amorphous characteristics, providing exceptional clarity and translucence. They have favorable flow characteristics for processing and are suitable for applications necessitating medical or food contact certifications. The mechanical qualities can be adjusted by altering the comonomer ratio, with styrene generally increasing hardness and MMA contributing flexibility.

MMA and styrene can undergo copolymerization via free radical polymerization to provide a copolymer comprising alternating units of both monomers. The resultant copolymer may be random or alternating, contingent upon the polymerization circumstances.

Key Properties

- S-MMA copolymers are recognized for their exceptional transparency, particularly in transparent ABS materials such as MABS.
- Mechanical Characteristics:
- The use of styrene often enhances the toughness of the copolymer.

- Incorporating MMA can enhance flexibility.
- Toughness: S-MMA copolymers demonstrate commendable toughness and impact resistance.
- They possess exceptional flow characteristics, rendering them appropriate for many processing techniques.
- MABS, a copolymer composed of MMA, styrene, acrylonitrile, and butadiene, exhibits excellent chemical resistance.
- S-MMA copolymers can preserve their characteristics post-sterilization.

Polyethyleneterephthalate (PET)

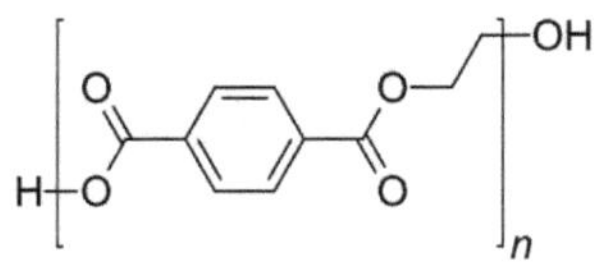

PET (Polyethylene Terephthalate) is a thermoplastic polymer belonging to the polyester category. This is a copolymer composed of terephthalic acid and ethylene glycol.

Key Properties

- PET is a robust and inflexible material, rendering it appropriate for applications necessitating structural integrity.
- It exhibits resistance to numerous substances, including acids, and can endure exposure to various solvents.
- PET exhibits low permeability to oxygen, carbon dioxide, and water vapor, rendering it suitable for packaging applications.
- PET demonstrates excellent electrical insulating characteristics, rendering it appropriate for application in electrical components.
- PET is a highly recyclable substance, enhancing its environmental sustainability.
- PET is a translucent substance, facilitating clear packaging and various applications.
- PET exhibits minimal moisture absorption, hence preserving its physical qualities and preventing degradation.
- PET can endure a variety of temperatures without considerable deterioration.
- PET is comparatively simple to mold and produce, rendering it appropriate for diverse manufacturing techniques.

Important Applications

Packaging

- Beverage containers: PET is extensively utilized for bottles, particularly for water, soda, and various beverages, owing to its durability, clarity, and recyclability.
- Food packaging: It is utilized for various food products, including trays, containers, and films, due to its barrier qualities.

Fabrics

- PET fibers are utilized in apparel, upholstery, and carpeting, providing durability, wrinkle resistance, and stain resistance.
- Sportswear: PET is favored for sportswear because of its rapid drying capabilities and its robust, lightweight characteristics.

Electronics and Additional Sectors

- PET is applicable in numerous electrical components, such as encapsulating materials and liquid crystal displays.
- Vehicle components: It is utilized in certain vehicle parts, including interior elements and insulation tape.
- Medical supplies: PET is employed in disposable surgical instruments and several medical applications.

Polyvinyl Chloride (PVC)

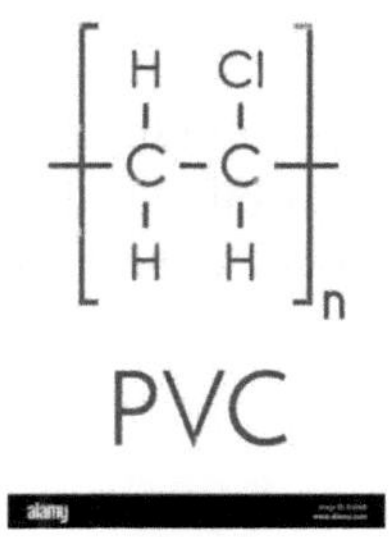

PVC (polyvinyl chloride) is a multifaceted thermoplastic recognized for its resilience, chemical resistance, and economic efficiency.

Key Properties

- PVC is recognized for its durability and resistance to abrasion.
- It exhibits resistance to a broad spectrum of chemicals, encompassing acids, alkalis, and saline solutions.
- PVC offers superior insulation for wires and cables, mitigating the risk of electrical shocks.

- PVC possesses intrinsic flame-retardant properties owing to its chlorine composition, rendering it a safer material for diverse uses.
- PVC is very economical in relation to other materials, rendering it a cost-effective option.
- PVC is readily malleable, allowing for shaping, molding, and welding, which facilitates many uses.
- PVC exhibits resistance to UV degradation and environmental wear, guaranteeing its durability in exterior applications.
- Notwithstanding its durability, PVC is quite lightweight, facilitating ease of handling and transportation.

Important applications

Construction and Architecture

- Pipes and Fittings: PVC pipes are extensively utilized in water, sewage, and drainage systems.
- PVC's resilience to weather and durability render it a favored option for window frames and external cladding.
- PVC is utilized in durable flooring and roofing membranes.
- PVC is utilized in wall coverings for both interior and exterior purposes.

Medical Services

- Medical Devices: PVC is utilized in blood bags, transfusion tubes, and surgical gloves.

Packaging

- PVC is utilized in the packaging of medical products to guarantee their sterility and safety.
- Flexible Packaging: PVC is utilized in the packaging of food, pharmaceuticals, and various products, offering flexibility and protection.
- PVC is suitable for the production of labels and presentation trays.
- Plastic Bottles: PVC is utilized in the manufacturing of plastic bottles for diverse liquids.

Electrical

- PVC serves as insulation for wires and cables, safeguarding them from electrical risks and environmental influences.
- PVC conduit serves to safeguard electrical wires and cables, enhancing their safety during installation and maintenance.

Automobile Industry

- PVC is utilized for insulation and sheathing in automotive wiring.
- PVC is utilized in underbody coatings to safeguard the vehicle's chassis from rust and corrosion.

Interior Trimmings

- PVC is utilized in interior trimmings, offering durability and visual appeal.

Alternative Applications

- PVC is utilized in apparel, including raincoats and aprons, offering waterproof and resilient materials.
- PVC is utilized in the manufacture of footwear, encompassing rubber alternatives and garden hoses.

Nylon 66 (PA66)

Nylon 66 is a robust and resilient thermoplastic recognized for its exceptional strength, rigidity, and resistance to wear and abrasion.

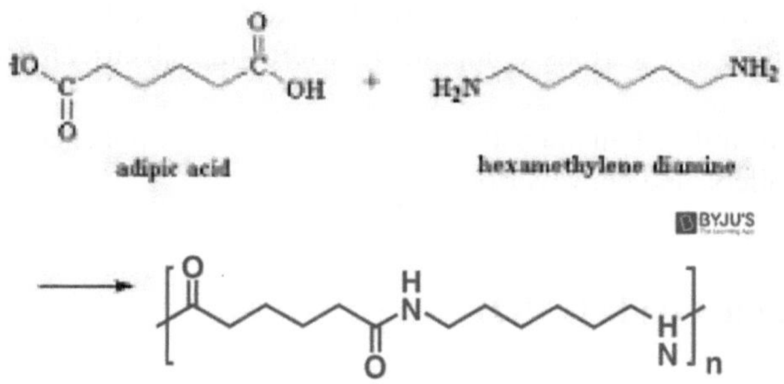

Structure of Nylon – 6,6

Key Properties

Mechanical Characteristics

- Nylon 66 may endure considerable tensile stresses before to failure.
- It withstands bending and distortion when subjected to load.
- It retains its form and withstands bending or distortion.
- It can endure abrupt shocks and impacts without breaking.
- It has exceptional resistance to abrasion and wear, rendering it appropriate for applications involving friction between components.
- It can endure repeated stress without fracturing or fissuring.

Thermal Characteristics

- Nylon 66 can endure elevated temperatures without compromising its structural integrity.
- Nylon 66 may retain its form and dimensions when subjected to thermal stress.
- It retains its form and measurements despite variations in temperature and humidity.

Chemical Characteristics

- It exhibits resistance to several substances, including fuels, oils, and alkalis.
- It exhibits resistance to a broad spectrum of hydrocarbon solvents.

Additional Properties

- Nylon 66 serves as an effective insulator, rendering it appropriate for electrical components.
- It possesses inherent self-lubricating characteristics, minimizing friction and wear.

Significant Applications

- Nylon 66 is a favored material for fibers in apparel, carpets, and many textile items because of its durability and abrasion resistance.
- Automotive: It is utilized in engine components, fuel systems, electrical connectors, and under-the-hood components due to its ability to endure heat, chemicals, and mechanical stress.
- Nylon 66's insulating characteristics render it appropriate for electrical connectors, switches, and various components within the electrical and electronics sector.
- Nylon 66 is utilized in industrial machinery for components such as gears, bearings, rollers, and conveyor parts.

Polyphenylene Oxide (PPO)

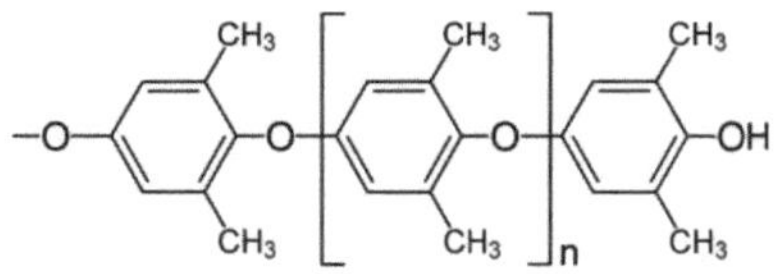

Key Properties

- Polyphenylene oxide (PPO), alternatively referred to as polyphenylene ether (PPE), is a high-performance thermoplastic material.

  The principal characteristics of PPO are as follows:

- PPO retains its form and dimensions despite stress and temperature fluctuations, rendering it appropriate for applications necessitating long-term dependability.
- The minimal moisture absorption of PPO enables it to maintain its electrical and mechanical qualities in humid conditions.
- PPO may endure elevated temperatures without considerable distortion or degradation of its characteristics.
- The low dielectric constant and low dielectric loss of PPO render it appropriate for electrical insulation applications.

- PPO can be rendered flame-resistant, rendering it appropriate for applications where fire safety is paramount.
- PPO has resistance to degradation by water and steam, rendering it appropriate for use in humid conditions.
- PPO exhibits an optimal equilibrium of strength and resilience.
- PPO demonstrates commendable flexural strength, indicating its capacity to endure bending loads.

Important Applications

- PPO is utilized in electrical connectors, circuit boards, and various components owing to its favorable electrical characteristics and dimensional stability.
- PPO's thermal resistance and durability render it appropriate for automotive components, including engine parts and interior trim.
- PPO is utilized in industrial applications requiring great resistance to heat, moisture, and chemicals.
- The biocompatibility and sterilization resistance of PPO render it appropriate for medical devices, including pump impellers and sterilization trays.
- PPO is utilized in domestic appliances, including pump impellers and diffusers, owing to its thermal resistance and endurance.

Polylactic Acid (PLA)

Polylactic acid (PLA) is a biodegradable thermoplastic polymer derived from renewable resources such as corn, sugarcane, or cassava.

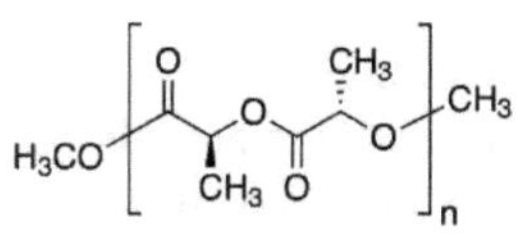

Key Properties

PLA is hydrolytically degradable and biocompatible, rendering it appropriate for biomedical applications.

Thermal Characteristics

- Glass Transition Temperature (T_g): Approximately 60 °C.
- Melting Temperature (T_m): 130–180 °C. Poly-L-lactic acid (PLLA) possesses a melting temperature (T_m) ranging from 170 to 180 °C.
- Heat Resistance: PLA endures temperatures up to 110 °C.

Mechanical Characteristics

- Tensile strength varies based on orientation and processing methods.
- Young's Modulus: 2.7–16 GPa.
- Elongation at Break: Up to 30% in orientated films.
- Rockwell Hardness: R70 to R90.

Additional Physical Properties

- Density: Amorphous PLA: 1.248 g/cm^3; Crystalline PLA: 1.290 g/cm^3.
- Solubility: Partially soluble in solvents such as acetone, toluene, and tetrahydrofuran (THF) at lower temperatures; fully soluble at boiling points.
- PLA is readily amenable to melt processing and orientation.

Important Applications

- 3D Printing: PLA is a widely utilized filament for 3D printing owing to its processing simplicity, minimal warping, and effective layer adhesion.
- Medical Applications: PLA is utilized in medical implants, drug delivery systems, sutures, and tissue engineering owing to its biocompatibility and biodegradability.
- The textile industry utilizes PLA fibers to produce a range of textiles, such as apparel, beds, and carpets.
- Food Packaging: PLA is utilized for the packaging of food and beverages, encompassing films, bags, and containers.
- Disposable Products: PLA is utilized in disposable cutlery, straws, and other single-use items, providing a biodegradable alternative.

Additional applications for PLA include:

- Agricultural films: Safeguarding crops and vegetation.
- Electronics: Enclosures and casings.
- Construction: Joinery and insulating materials.
- Automotive components: Specific elements of automobiles.

Poly Hydroxy Alkanoates (PHA)

Polyhydroxyalkanoates (PHAs) are bioplastics possessing a variety of advantageous characteristics. They are biodegradable, biocompatible, and exhibit thermoplastic properties akin to synthetic plastics. PHAs have flexibility, elasticity, and good thermal stability.

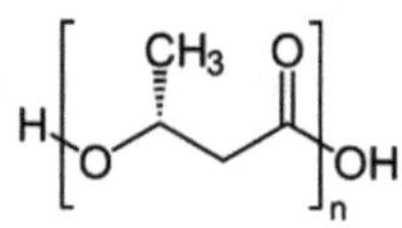

Key Properties

Physical and Chemical Characteristics

- Biodegradability: PHAs are decomposable by microorganisms in many conditions, rendering them a sustainable substitute for petroleum-derived polymers.
- Biocompatibility: PHAs are often well-accepted by the body, rendering them appropriate for biomedical purposes.
- Thermoplasticity: Polyhydroxyalkanoates (PHAs) can be manipulated utilizing standard plastic processing techniques such as injection molding and thermoforming.
- Crystallinity: PHAs may exhibit semi-crystalline characteristics, with differing levels of crystallinity influencing their properties.
- Flexibility: Polyhydroxyalkanoates (PHAs) may exhibit either flexibility or rigidity based on their monomer composition.
- Transparency/Colorability: PHAs may exhibit transparency, translucency, or coloration based on the specific application.
- Solubility: PHAs are typically insoluble in water and organic solvents, although soluble in halogenated solvents such as chloroform.
- Barrier Characteristics: PHAs have excellent barrier qualities against gases and oxygen, rendering them appropriate for packaging applications.
- The density of PHA fluctuates according on its chemical makeup, generally ranging from 1 to 1.3 g/cm^3.
- Melting Point: PHAs have melting values between 40 and 180 °C.

Mechanical Characteristics

- Flexibility and Elasticity: PHAs may possess flexibility and elasticity, with certain types demonstrating mechanical qualities akin to those of synthetic plastics.
- Tensile strength and Young's modulus: PHAs exhibit tensile strength and Young's modulus comparable to synthetic polymers.
- Ductility and Impact Resistance: The composition of PHAs can render them ductile with commendable impact resistance.

Additional Significant Attributes

- Processability: PHAs can be utilized with standard plastic processing machinery.
- UV Stability: PHAs have a high degree of UV stability.
- Water Resistance: PHAs exhibit limited water permeability.
- Elevated Temperature Stability: Certain PHAs, such as PHBV and PHBH, have commendable stability at high temperatures, rendering them appropriate for packaging applications.

Important Applications

- Gentle Exfoliation: PHAs effectively eliminate the outermost layer of dead skin cells, uncovering a more radiant and smoother complexion.
- Hydration and Moisture: PHAs function as humectants, attracting moisture to the skin, hence maintaining hydration and a plump appearance.

- Anti-Aging Advantages: By promoting cellular turnover and enhancing skin texture, PHAs can diminish the visibility of fine lines and wrinkles.
- Enhanced Skin Barrier: PHAs can fortify the skin barrier, increasing its resilience against environmental damage and irritation.
- Appropriate for Sensitive Skin: Owing to their milder characteristics, PHAs are frequently advised for individuals with sensitive or dry skin who may not endure other acids.
- Applicable for Diverse Dermatological Issues: PHAs can enhance skin tone and texture, diminish hyperpigmentation, and lessen the visibility of acne and outbreaks.

Polyglycolic Acid (PGA)

Commonly referred to as polyglycolide, it is a biodegradable and biocompatible polymer possessing distinctive features. It demonstrates elevated crystallinity, resulting in superior mechanical strength, enhanced gas barrier characteristics, and solubility in specific solvents.

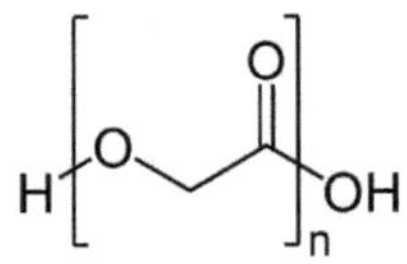

Key Properties

- PGA is a biodegradable polymer, indicating that it degrades within the body over time by hydrolysis.
- The elevated crystallinity enhances fiber strength and modulus.
- PGA is recognized for its superior gas barrier qualities, advantageous in specific applications like as packaging.
- High molecular weight PGA is insoluble in most conventional organic solvents, although lower molecular weight oligomers have greater solubility. It is soluble in strongly fluorinated liquids.
- PGA possesses a high melting point of around 225 °C and a glass transition temperature ranging from 35 to 40 °C.
- PGA is well-accepted by the body, eliciting minimal tissue response, rendering it appropriate for medical purposes.
- PGA is recognized for its superior fiber-forming properties, rendering it appropriate for applications such as sutures.

Important Applications

Applications in Medicine and Biomedical Fields

- Surgical Sutures: PGA is a prevalent substance for absorbable sutures, utilized to close wounds and approximation tissues throughout surgical procedures.
- Tissue Engineering: PGA serves as a scaffold material for tissue regeneration in domains such as bone, cartilage, and spinal regeneration.
- Drug Delivery Systems: PGA and its copolymers are utilized to fabricate biodegradable drug delivery devices, facilitating the regulated release of pharmaceuticals.
- PGA is utilized in the fabrication of diverse implantable medical devices, including screws, stents, and plates, for orthopedic and reconstructive surgical procedures.
- Orthopedic Applications: PGA is utilized in the fabrication of implants for fractures, bone regeneration, and various orthopedic ailments.

Alternative Applications

- The gas barrier qualities of PGA render it appropriate for the production of multilayer films and packaging that safeguard food from deterioration and flavor degradation.
- PGA is investigated for utilization in oil drilling instruments, gas-barrier bottles, and several industrial applications.

Essential Attributes Enhancing its Versatility

- Biodegradability: PGA decomposes naturally within the body, eliminating the necessity for the extraction of implants or sutures.
- Biocompatibility: PGA is highly tolerated by the body, reducing inflammation and other negative responses.
- PGA fibers demonstrate a high Young's modulus, rendering them appropriate for structural support in tissue engineering and many applications.
- PGA possesses gas barrier properties, rendering it appropriate for food packaging and other applications requiring controlled gas permeability.

Polyamide 11 (PA11)

PA11, or Polyamide 11, is a bioplastic sourced from castor oil, characterized by exceptional ductility, impact resistance, and chemical resistance.

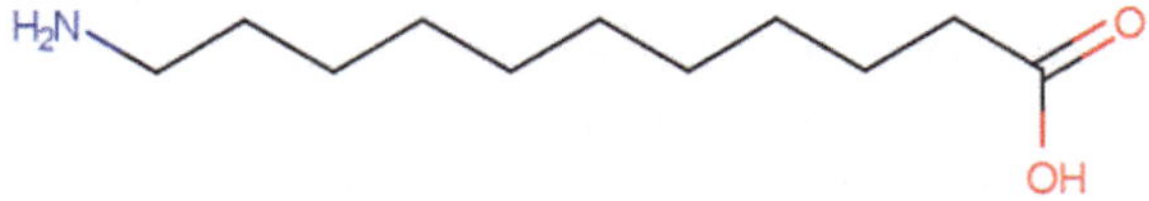

Key Properties

Mechanical Characteristics

- Tensile Strength: 47–58 MPa (about 7–8 ksi).
- Tensile Modulus: 1560–1700 MPa (about 226,000–247,000 psi).
- Elongation at Break: 40–51%.
- Flexural Modulus: 1200 MPa (about 174,000 psi).
- Impact Resistance: Elevated, exhibiting considerable impact strength for a bioplastic.
- Hardness: 75–80 Shore D.

Thermal Characteristics

- Melting Point: 185–200 °C (365–392°F).
- Heat Deflection Temperature at 0.45 MPa: 47 °C (117°F).
- Heat Deflection Temperature (1.82 MPa): 185 °C (365°F).
- Coefficient of Thermal Expansion: 171 (0.095) μm/°C m (thou/in·°F).

Additional Properties

- Density: 1.03–1.04 g/cm^3.
- Water Absorption: Minimal, rendering it appropriate for use in humid conditions.
- Chemical Resistance: Excellent resistance to chemicals, ultraviolet light, and severe weather conditions.
- Smooth Surface Finish: Produces a low coefficient of friction, appropriate for situations where surface friction is a consideration.
- Permeability: Exhibits minimal permeability to hydrocarbons.
- Environmentally Sustainable: Sourced from renewable materials such as castor oil.

Important Applications

Automobile Sector

- Fuel lines, brake lines, and air conditioning systems: PA11's robustness and chemical resistance render it appropriate for these applications.
- Clamps, brackets, connectors, and wire forms: PA11's resistance to corrosion and scratches, along with its sound-dampening properties and aesthetic enhancement, render it appropriate for various applications.
- Bearings for windshield wiper arms and various automotive components: The mechanical qualities of PA11 render it an appropriate material for these components.
- Athletic Equipment: PA11 coatings on sporting equipment necessitate minimal maintenance and improve their aesthetic appeal.
- PA11 is utilized in badminton shuttlecocks, ski top layers, racket strings, and eyelets because of its resistance to moisture, corrosion, and low density.
- Sports shoe soles: PA11 is utilized in the soles and various mechanical components of footwear.

Additional Sectors

- Electronics: PA11's resistance to electric arcs and electrolytic corrosion renders it optimal for wire and cable insulation, hence mitigating the danger of electrical failure.
- Medical equipment: The biocompatibility and chemical resistance of PA11 render it appropriate for catheters, tubing, and various medical equipment.
- Textiles and Consumer Goods: PA11 is utilized in fabrics and fibers for athletic apparel, protection equipment, and resilient garments.
- Energy and Environmental Technology: PA11 is optimal for flexible pipelines, cable insulation, and renewable energy systems.
- Pipes for natural gas conveyance: The strength and durability of PA11 render it appropriate for this use.
- Automotive hoses: The flexibility and fluid resistance of PA11 render it appropriate for this use.
- 3D Printing: PA11 is applicable in 3D printing for functional prototypes and limited production runs.
- Precision moldings for engineering applications: PA11 is appropriate for this purpose owing to its favorable mechanical qualities.
- Travel bags: PA11 is utilized in travel bags due to its toughness and resistance.

Phenol Formaldehyde Resin

Known as phenoplasts, these thermosetting polymers demonstrate characteristics such as effective thermal insulation, low density, exceptional durability, and chemical resistance.

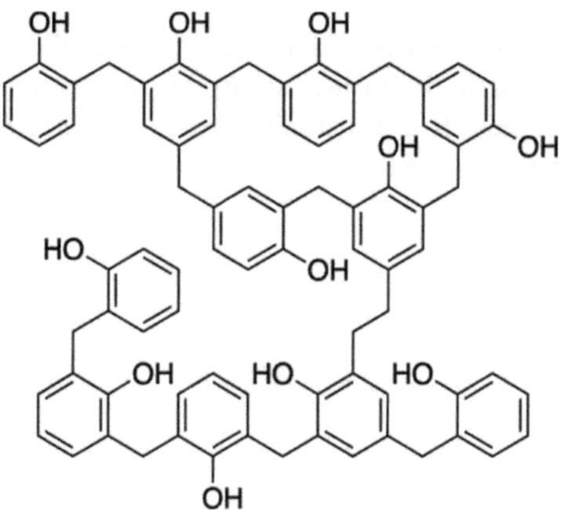

Key Properties

- Excellent Thermal Insulation: Phenol-formaldehyde resins exhibit significant heat resistance and can endure elevated temperatures, rendering them appropriate for uses such as electrical insulation and heat-resistant components.
- Low Density: In comparison to other resins such as epoxy, phenol-formaldehyde exhibits a lighter weight, facilitating ease of handling and rendering it appropriate for applications where weight is a critical consideration.
- Exceptional Durability: Phenoplasts exhibit resilience and resistance to degradation, providing substantial strength and lifespan.
- Excellent Chemical Resistance: They exhibit resistance to numerous chemicals and solvents, rendering them appropriate for applications where chemical exposure poses a risk. Nonetheless, they are inappropriate for settings with strong acids or alkalis.
- Moisture Resistance: Phenol-formaldehyde resins exhibit notable moisture resistance, rendering them appropriate for applications such as outdoor plywood.
- Their superior electrical insulating characteristics render them appropriate for electrical and electronic components.
- Bonding: They may establish chemical connections with wood constituents, rendering them advantageous in adhesives and wood-derived goods.

Categories of Phenol-Formaldehyde Resins

- Novolac Resins: Acid-catalyzed and necessitate a curing agent for solidification.
- Resole Resins: Alkaline-catalyzed and cured without a curing agent.

Urea Formaldehyde Resin

This thermoset resin demonstrates various significant qualities, including elevated tensile and flexural strength, little water absorption, and commendable heat distortion temperature tolerance. It also exhibits elevated surface hardness, volumetric resistance, and elongation at fracture.

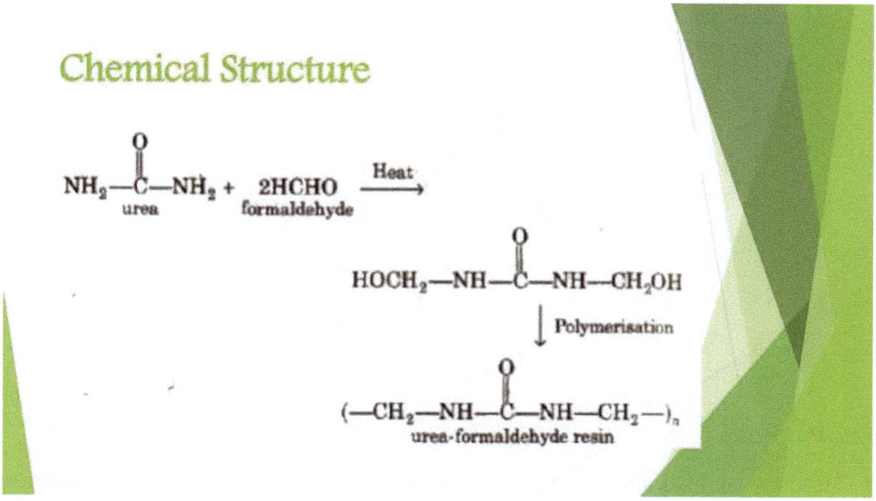

Key Properties

- Elevated Tensile Strength: UF resin endures considerable stress when subjected to tension prior to failure.
- Elevated Flexural Modulus: It can withstand bending deformation when subjected to stress.
- Elevated Heat-Distortion Temperature: Prevents softening or deformation under high temperatures.
- Minimal Water Absorption: Impervious to moisture infiltration.
- Mold shrinking: Exhibits negligible shrinking throughout the curing process.
- Elevated Surface Hardness: Ensures a resilient and scratch-resistant exterior.
- Elongation at Break: The capacity to stretch or extend before to fracture.
- Volume Resistance: Opposes the flow of electrical electricity.
- Thermosetting Property: Achieves permanence upon heating and curing.
- Viscosity: The liquid state of UF resin has a certain viscosity, generally ranging from 350 to 600 mPa s at 20 °C.
- The density of UF resin is approximately 1280–1290 kg/m^3 at 20 °C.
- Refractive Index: It possesses a refractive index of 1.55.

Melamine Formaldehyde Resin

It is a thermosetting polymer recognized for its hardness, durability, and resistance to heat, fire, and chemicals. It is a multifaceted material employed in numerous applications owing to its excellent mechanical, thermal, and electrical characteristics.

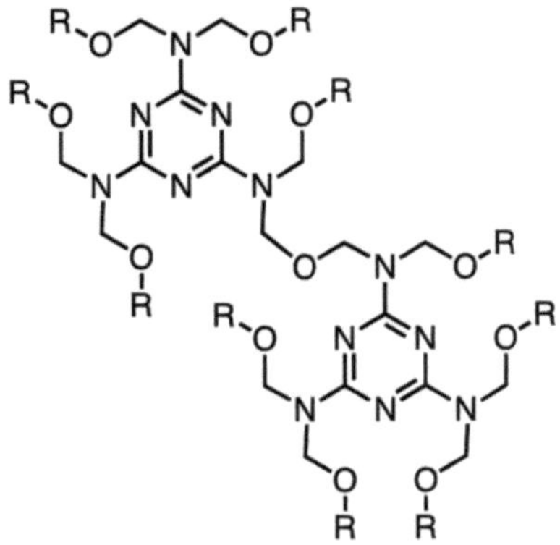

Key Properties

- Hardness and Durability: MF is an exceptionally durable and rigid thermosetting aminoplast.
- Heat and Fire Resistance: It demonstrates remarkable resistance to heat and fire, mostly owing to its capacity to emit nitrogen gas upon combustion, which aids in extinguishing flames.
- Cured MF polymers have chemical resistance and hydrolytic stability, rendering them appropriate for many applications.
- Favorable Mechanical Properties: MF provides substantial tensile strength, flexural modulus, and impact resistance.
- Electrical Insulation: It possesses excellent electrical insulation characteristics, including resistance to tracking.
- Transparency: In certain applications, MF may exhibit transparency.
- Flame Retardant: Melamine derivatives have flame retardant qualities owing to their nitrogen concentration.
- Moisture Resistance: MF exhibits excellent moisture resistance, rendering it appropriate for outdoor and wet conditions.
- Surface Smoothness: MF is recognized for its polished surface quality.
- Boil Resistance: MF exhibits resistance to boiling water, rendering it appropriate for food contact applications.
- Abrasion Resistance: MF demonstrates excellent resistance to abrasion, guaranteeing prolonged durability.
- Scratch Resistance: MF exhibits resistance to scratches, preserving its aesthetic and operational integrity.
- Favorable Electrical Characteristics: Cured MF materials have excellent electrical conductivity and insulating qualities.

Important Applications

- MF resins are extensively utilized as adhesives across diverse applications, such as wood bonding, laminates, and coatings.
- MF coatings are utilized in diverse applications, such as fire-retardant coatings, paints, and finishes.
- Melamine formaldehyde is an essential element in ornamental and high-pressure laminates utilized in flooring, worktops, and furniture.
- MF is utilized to manufacture molded plastic items, such as dinnerware, buttons, and electrical components.
- MF's hardness, durability, and safety for food contact render it appropriate for dinnerware and other food-related items.
- The electrical insulating qualities of MF render it appropriate for wiring devices and various electrical components.

Epoxy Resin

Epoxy resin is known for its strong adhesion, resistance to chemicals and moisture, and high strength. It also exhibits excellent electrical insulation, heat resistance, and low shrinkage during curing.

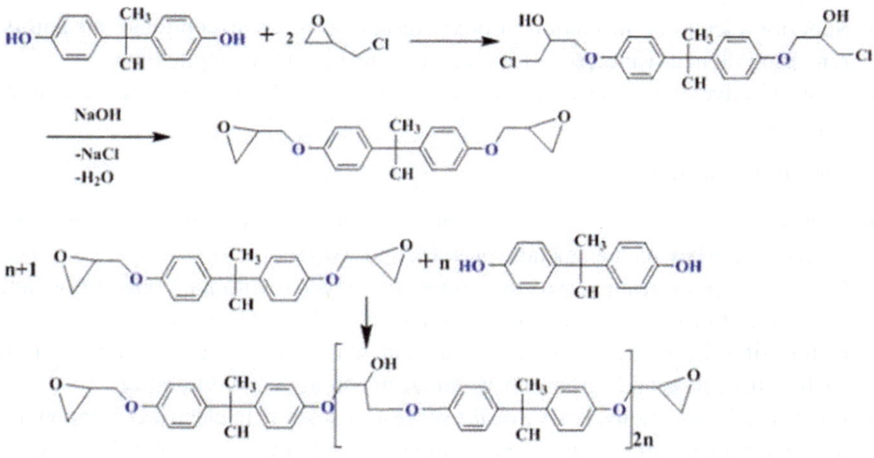

Key Properties

Mechanical Characteristics

- Epoxy resins have exceptional tensile, compressive, and impact strength, rendering them appropriate for structural applications.
- Good toughness: Although epoxy resins may exhibit brittleness, their flexibility and toughness can be enhanced through methods such as rubber toughening.
- Little shrinkage: The little shrinkage during curing is beneficial, since it reduces distortion and tension in the final product.

Resistance to Chemicals and Environmental Factors

- Chemical resistance: Epoxy resins exhibit resistance to a diverse array of chemicals, encompassing acids, bases, solvents, and oils.
- Moisture resistance: They exhibit significant resistance to moisture, rendering them appropriate for applications subjected to water or humidity.
- Thermal resistance: Epoxy resins endure moderate temperatures, with certain formulations capable of withstanding elevated temperatures.

Additional Significant Attributes

- Superior adhesion: Epoxy resins exhibit strong adherence to diverse substrates, such as metals, concrete, and wood, rendering them optimal for bonding and coating purposes.
- Superior electrical insulation: Epoxy resins serve as exceptional insulators, rendering them advantageous for electrical and electronic applications.
- Cost-effective: Epoxy resins are typically more affordable than several other high-performance materials.

Important Applications

- Adhesives: Epoxy resins are robust and resilient adhesives utilized for attaching diverse materials such as metals, plastics, and wood.
- Coatings: Epoxy resin serves as a protective layer for floors, countertops, and furniture, offering resistance to abrasion, chemicals, and moisture.
- Composites: Epoxy resins serve as a fundamental element in composite materials such as fiberglass and carbon fiber, imparting strength and durability.
- Electrical Applications: Epoxy resins serve as insulators in electrical components and electronics, safeguarding against moisture and environmental pollutants.
- Building and Repairs: Epoxy resin is utilized in building for the remediation of concrete fissures, adhesion of metal and wood, and various other repairs.
- Marine and vehicle: Epoxy resin is utilized in these sectors for hull construction, boat fabrication, and vehicle body repairs.
- Creative Arts: Epoxy resin is utilized in artistic endeavors for the production of resin paintings, jewelry, and many decorative objects.
- DIY Projects: Epoxy resin serves as a multifaceted material for an array of DIY endeavors, encompassing furniture, countertop treatments, and bonding or repair applications.

Polyurethane (PU)

These resins are multifaceted polymers exhibiting a broad spectrum of attributes, including elevated strength, durability, and resistance to chemicals and moisture.

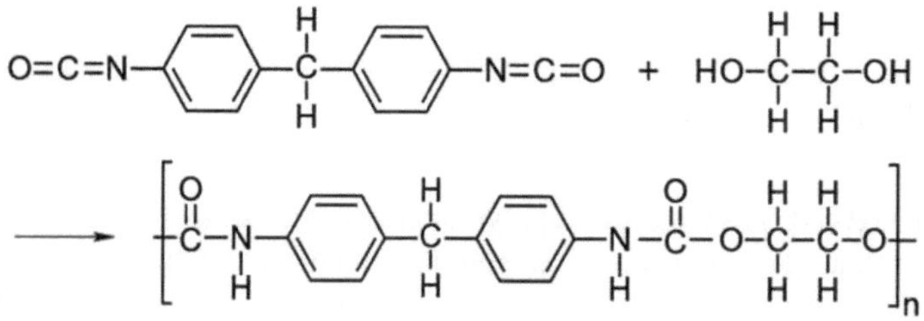

Key Properties

- Exceptional Strength and Durability: PU resins are recognized for their resilience and capacity to endure mechanical stress and shocks.
- Chemical Resistance: They exhibit resistance to several chemicals, solvents, and water, rendering them appropriate for demanding situations.
- Water Resistance: PU resins exhibit resistance to water and moisture, rendering them suitable for outdoor applications and environments with elevated humidity.
- Flexibility and Elasticity: PU resins can be engineered to exhibit significant flexibility and elasticity, rendering them appropriate for use in foams and elastomers.
- Superior Insulating Properties: Certain PU resins, especially foams, have exceptional insulating characteristics, rendering them appropriate for insulation and thermal management applications.
- Excellent Adhesion: PU resins can establish robust connections with diverse materials, rendering them advantageous in adhesives and coatings.

Important Applications

- Coatings: Polyurethane resins serve as protective coatings for diverse surfaces, ensuring durability in automotive, industrial, and marine applications.
- Adhesives and Sealants: Polyurethane resins are utilized to formulate robust and enduring adhesives and sealants for the adhesion of diverse materials, such as wood, metal, and plastic.
- Elastomers: Polyurethane resins are utilized to produce elastomers, which are pliable and durable materials appropriate for applications such as tires, shoe bottoms, and shock absorbers.
- Polyurethane foams primarily consist of PU resins and are utilized in many applications such as cushioning, insulation, and packaging.
- PU resins are utilized in construction for insulation, roofing, and coatings.
- Vehicle: Polyurethane resins are utilized in vehicle components, such as bumpers, interior elements, and engine parts.

- Electronics: PU resins are utilized in electronics for encapsulating components, offering insulation, and improving durability.
- Footwear: Polyurethane resins are utilized in footwear for soles and various components, offering comfort and durability.
- PU resins are utilized in furniture for cushioning, upholstery, and coatings.

Polyaniline

Polyaniline (PANI) is a conductive polymer characterized by diverse characteristics and applications. It demonstrates exceptional electrical conductivity, optical characteristics, and thermal stability, rendering it appropriate for various applications, including energy storage, sensing, and electronic devices.

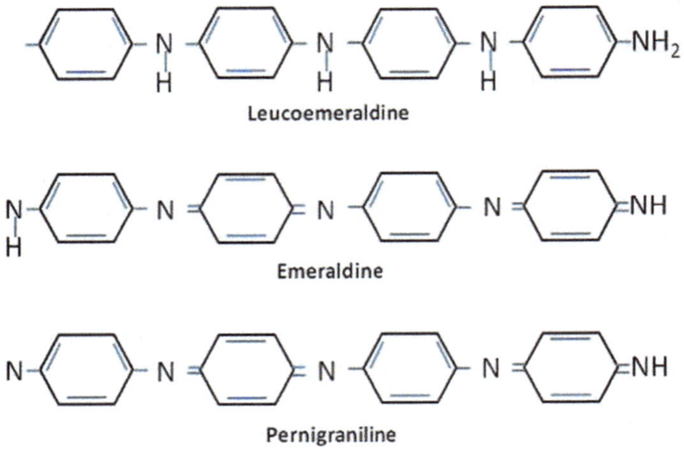

Key Properties

- The elevated electronic conductivity of PANI arises from its π-conjugated electron system. It can be doped with diverse materials to augment its conductivity.
- The optical properties of PANI vary with its oxidation state, enabling its application in electrochromic devices and display technologies.
- Thermal Stability: PANI has resistance to elevated temperatures, rendering it appropriate for diverse applications, including those requiring high-temperature processing.
- Chemical Stability: PANI has considerable stability in both acidic and alkaline conditions, ensuring an extended lifespan across diverse applications.
- The mechanical characteristics of PANI can fluctuate based on its synthesis technique and the inclusion of additives. For instance, solution-prepared PANI exhibits greater rigidity and superior mechanical properties compared to PANI synthesized using alternative methods.

Important Applications

- Energy Storage: PANI is used in supercapacitors and rechargeable batteries owing to its elevated specific capacitance and commendable cycling stability.
- PANI is utilized in gas sensors (such as ammonia and carbon monoxide detection), humidity sensors, and biosensors.
- PANI is applicable in organic field-effect transistors (OFETs), organic light-emitting diodes (OLEDs), and many organic electronic devices.
- Optoelectronic Devices: The optical characteristics of PANI render it appropriate for application in photonic and optoelectronic devices.
- PANI is utilized in anti-corrosion coatings owing to its capacity to inhibit corrosion and its environmental resilience.
- PANI-based adsorbents are effective in eliminating pollutants from water, such as heavy metals and organic contaminants.
- PANI is utilized in electroactive inks, paints, coatings, adhesives, and conductive fibers.

Challenges

- Processability: PANI exhibits suboptimal processability and mechanical qualities, constraining its applicability in actual scenarios.
- Solubility: PANI exhibits low solubility in aqueous environments owing to its hydrophobic characteristics, which may hinder stable film formation.
- Mechanical Strength: Polyaniline (PANI) exhibits low mechanical strength, which can be enhanced through its integration into nanocomposites.

Polythiophene

Polythiophene is a semiconducting conjugated polymer recognized for its superior electrical conductivity, thermal and environmental resilience, and excellent processability.

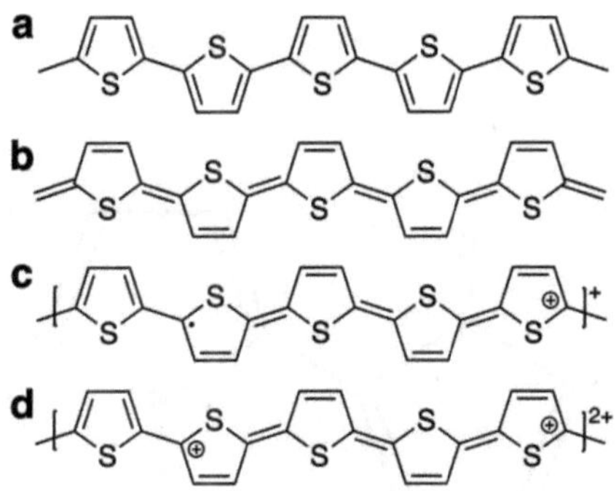

Key Properties

- Electrical Conductivity: Polythiophene has excellent electrical conductivity, particularly when doped.
- Thermal and Environmental Stability: It demonstrates significant thermal and environmental stability, rendering it appropriate for diverse applications.
- Processability: Polythiophene can be readily fabricated into several forms, including films, fibers, and nanoparticles.
- Optical Properties: It possesses distinctive optical characteristics, including photoluminescence, which can be adjusted via doping or chemical alteration.
- Solubility: Unsubstituted polythiophenes exhibit low solubility in conventional solvents, although soluble derivatives such as polyalkylthiophenes are easily obtainable.
- Doping: Polythiophene is readily dopable, markedly improving its electrical conductivity.

Important Applications

- Energy Storage: Utilized in supercapacitors, rechargeable batteries, and various energy storage systems.
- Optoelectronics: Utilized in polymer light-emitting diodes (LEDs), photovoltaic cells, and optical sensors.
- Sensors: Employed in chemical and optical sensors, biosensors, and many sensing applications.
- Nonlinear Optics: Employed in nonlinear optical devices owing to its distinctive optical characteristics.
- Antistatic Coatings: Employed as antistatic coatings owing to their electrical conductivity.
- Additional Applications: Utilized in photoresists, anti-corrosion coatings, and various other uses.
- Biomedical Applications: Prospective uses in pharmacological delivery, tissue engineering, and biosensing technologies.

Polypyrrole

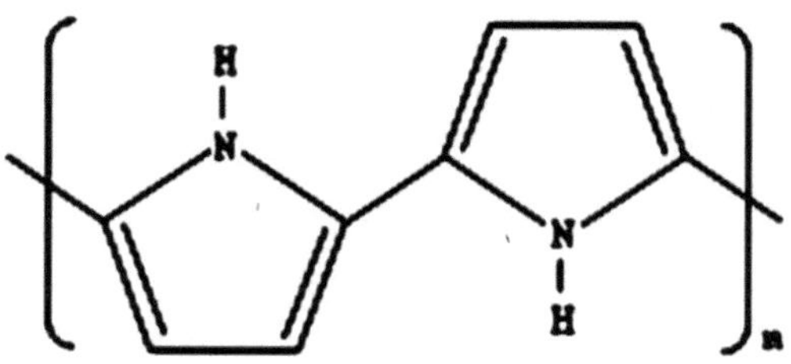

Key Properties

- PPy is an inherently conductive polymer with satisfactory electrical conductivity.
- It possesses favorable mechanical properties, including tensile strength and flexibility.
- PPy has chemical stability in both air and water, advantageous for specific applications.
- PPy's biocompatibility renders it appropriate for biomedical applications, including biosensors and drug delivery systems.

Important Applications

- PPy is utilized in supercapacitors, batteries (including fuel cells, dye-sensitized solar cells, and lithium/sodium batteries), and various energy storage technologies.
- Polypyrrole (PPy) is utilized in diverse sensor applications, encompassing gas sensors, biosensors, and chemical sensors.
- Polypyrrole (PPy) is utilized in drug delivery systems, tissue engineering, and several biomedical applications.
- PPy is utilized in water pollution management technologies, encompassing separation, purification, oxidation, and disinfection.
- PPy serves as a coating material for corrosion protection.
- PPy is utilized in electronics, optical devices, and various applications, including computer screens and field-effect transistors.
- PPy serves as a shielding material to safeguard against electromagnetic interference.
- Polypyrrole (PPy) can be utilized in actuators, including those employed in prosthetic devices.

Appendix B
Glass Transition Temperature and Melting Temperature of Polymers

Polymer name	T_g (°C)	T_m (°C)
ABS—Acrylonitrile butadiene styrene	90.0	190–220
CA—Cellulose acetate	100.0	230–300
EVOH—Ethylene vinyl alcohol	15.0	156–195
HDPE—High density polyethylene	−110.0	137
HIPS—High impact polystyrene	88.0	180–250
LDPE—Low density polyethylene	−110.0	137
LLDPE—Linear low density polyethylene	−110.0	120–130
PA 11	35.0	180–190
PA 12	35.0	175–180
PA 6–Polyamide 6	60.0	230–280
PA 66–Polyamide 6-6	55.0	265
PAI—Polyamide-imide	275.0	357
PBT—Polybutylene terephthalate	55.0	223
PET—Polyethylene terephthalate	73.0	270
PI—Polyimide	250.0	247–388
PLA, Injection molding	55.0	150–180
PMMA—Polymethylmethacrylate/acrylic	90.0	200
PP (Polypropylene) homopolymer	−10.0	160
PS, High heat	90.0	240
PVC Rigid	60.0	273
Natural rubber	−73	28

S. D. Sadhu et al., *Source Book of Experimental Techniques and Analysis of Polymeric Materials*, https://doi.org/10.1007/978-981-92-2088-5

Appendix C
List of Polymers, Their Solvents and Non-solvents

[Ref: http://kinampark.com/PL/files/Solubility%20Polymers%202011%2C%20Encyclopedia%20pf%20Polymerr%20Science%20and%20Technology.pdf]

Polymer	Solvent
Polystyrene	$SbCl_3$, $AsCl_3$, cyclohexane (above 35 °C), benzene, ethylbenzene, CH_2Cl_2, $CHCl_3$, chlorobenzene, cyclohexanone, methyl ethyl ketone, dimethyl formamide, ethyl acetate, methyl benzoate, 1,3-dioxolane, methyl to butyl phthalate, 1-nitropropane, pyridine CS_2
PMMA	Acetone, ethyl acetate, ethylene-chloride, ethyl glycol, amyl acetate, chlorobenzene, cyclohexanone, benzene, butyl acetate, dioxane, methyl acetate, butan-2-one, methylene chloride, methyl glycol, nitroethane, toluene, trichloroethylene, acetic acid
PVC	$SbCl_3$, cyclohexanone, cyclopentanone, diisopropylketone, butan-2-one, isophorone, dichlorobenzene, trichlorobenzene, dimethylformamide, mesityl oxide, nitrobenzene, tetrahydrofuran, hexamethyl phosphoric acid triamide, tricresyl phosphate, mixtures of acetone
PEG	Acetonitrile, ethylene carbonate, ethylene dichloride, benzene, chloroform, dimethylformamide, methanol, butanone, dichloromethane, carbon tetrachloride, tetrahydrofuran, trichloroethylene, water (cold); at slightly elevated temperatures also in acetone,
Polyacrylonitrile	Hexanedinitrile, ethylene carbonate, $SbCl_3$, gamma-butyrolactone, epsilon-caprolactam, cyanoacetic acid, dimethyl acetamide, dimethylformamide, dimethyl sulfone, dimethyl sulfoxide, hexafluoroacetone sesquihydrate, hydroxyacetonitrile, propanedinitrile

(continued)

S. D. Sadhu et al., *Source Book of Experimental Techniques and Analysis of Polymeric Materials*, https://doi.org/10.1007/978-981-92-2088-5

(continued)

Polymer	Solvent
PVA	Diethylenetriamine (cold), triethylenediamine, triethylenetetramine (cold), dimethyl sulfoxide (hot) glycols (hot), glycerol, hexafluoroacetone sesquihydrate, water, formamide, dimethylformamide
HDPE	Soluble only above 80 °C in aliphatic, aromatic, and cycloaliphatic hydrocarbons (e.g., decane p-xylene, decahydronaphthalene), halogenated aliphatic, aromatic, and cycloaliphatic hydrocarbons (e.g., heptylchloride, dichlorobenzene, chlorocyclohexane)
LDPE	Like HDPE, the temperature of solubility lies normally 20–30 °C lower depending on the degree of branching
Cellulose	Concentrated aqueous solutions of HF, HCl, HBr, H_2SO_4, H_3PO_4, $ZnCl_2$, $Ca(SCN)_2$ (always with degradation); aqueous solution of tetraethylammonium hydroxide, benzyltrimethylammonium hydroxide, dimethyldibenzylammonium hydroxide, tetraalkyphosphonium hydroxide
Polyhexamethylene adipamide (Nylon-6:6)	HCOOH, chloroacetic acid, HF, H_3PO_4, H_2SO_4, phenols, trichloroethanol, trifluoroethanol, chloral hydrate, SO_2, methanolic HCl-, $CaCl_2$-, and $MgCl_2$-solutions; furthermore between 120 and 180 °C: ethylene chlorohydrin, N-acetylmorpholine, benzyl alcohol
Poly-e-caprolactam (Nylon-6)	Formic acid, $SbCl_3$, $AsCl_3$, ethylene carbonate, benzyl alcohol, g-butyrolactone, chlorinated acetic acid, chlorophenol, dimethyl formamide, dimethyl sulfoxide, formamide, hexafluoroacetone sesquihydrate, phenethyl alcohol, phosphoric acid trisdimethyl amide
Poly-e-caprolactone	Acetone, aromatic hydrocarbons, ethyl acetate, ethylene dichloride, tetrachloroethane, carbon tetrachloride
Poly(vinyl acetate)	Benzene, toluene, halogenated aliphatic, and aromatic hydrocarbons, methyl, ethyl, and butyl acetate, acetone, methyl ethyl ketone, 4-methyl-2-pentanone, methanol, allyl alcohol, tetrahydrofuran, dioxane, glacial acetic acid, dimethyl formamide, dimethyl
Polyisobutylene and polybutadiene	Aliphatic, aromatic, and cycloaliphatic hydrocarbons (e.g., heptane, diisobutylene, toluene, tetrahydronaphthalene), chlorinated aliphatic, aromatic, and cycloaliphatic hydrocarbons (e.g., chloroform, CCl_4, dichlorobenzene), tetrahydrofuran, dioxane

MIX
Papier aus verantwortungsvollen Quellen
Paper from responsible sources
FSC® C105338

If you have any concerns about our products,
you can contact us on
ProductSafety@springernature.com

In case Publisher is established outside the EU,
the EU authorized representative is:
Springer Nature Customer Service Center GmbH
Europaplatz 3, 69115 Heidelberg, Germany

Printed by Libri Plureos GmbH
in Hamburg, Germany